Samuel John Hunter

The Coccidae of Kansas

A series of articles dealing with Coccidae found in Kansas, their host plants, and

bibliography

Samuel John Hunter

The Coccidae of Kansas
A series of articles dealing with Coccidae found in Kansas, their host plants, and bibliography

ISBN/EAN: 9783337300722

Printed in Europe, USA, Canada, Australia, Japan

Cover: Foto ©berggeist007 / pixelio.de

More available books at **www.hansebooks.com**

THE COCCIDÆ OF KANSAS.

A Series of Articles Dealing with
Coccidæ found in Kansas,
their Host Plants, and
Bibliography.

BY

S. J. HUNTER.

ASSOCIATE PROFESSOR IN THE UNIVERSITY OF KANSAS.

WITH TWENTY PLATES, COMPOSED OF
SEVENTY-TWO ILLUSTRATIONS.

PREFACE.

In preparing the series of articles which follow, the author has had in mind the needs of the student beginning a systematic study of the indigenous representatives of the Coccidæ. Accordingly, to facilitate their use, these papers are now brought together under one cover, and are preceded by a brief introduction. For reference and explanation of terms, an index and glossary are added. Since the purpose of this work has been to treat of species occurring in a state of nature in this latitude, but little attention has been given to those forms dwelling in greenhouses.

In the study of this group, the student is urged to give special thought and care to technique of the microscopical preparations. Obviously, well-prepared mounts are essential to proper anatomical studies. The student should not be satisfied with simple determinations of species, but should continue the study of the individual representatives of each species, brought together in many cases from different localities and on different hosts, in order to comprehend the variations and gradations existing within the species.

INTRODUCTION.

Origin of word. Gr. Kokkos, berry; specifically, the berry that grows upon the scarlet tree.

Distinguishing Characteristics. The Coccidæ, or scale insects, belong to the suborder Homoptera of the order Hemiptera. The Hemiptera are distinguished from insects of other orders by the presence of sucking mouth-parts, and an incomplete metamorphosis. Many of the Hemiptera possess four wings. The Coccidæ, anomalous as they are, conform strictly to none of these requirements. The males have two transparent wings, and are without mouth or food-taking apparatus. The natural position of the mouth is occupied by a pair of supplementary eyes. Instead of a second pair of wings there exists on each side of the metathorax a small hooked appendage. These, in some species, serve as hooks to attach or control the true wings. The males pass through a complete metamorphosis.

The female. The males and females are similar in form and structure when hatched. (See Plate V, fig. 23, and Plate XVI, fig. 3.) These mite-like creatures differentiate during growth and development. The females pass through an incomplete metamorphosis, in most cases losing the power of locomotion and becoming stationary bodies engaged in taking nourishment from plant tissues, and in reproduction. The life histories and habits, on account of the great variation, can be best discussed under the subfamilies.

Scale insects differ from closely related forms in the following particulars:

The legs in both sexes terminate in a single claw.

The females are wingless.

The adult males, with the exception of two or three species, possess one pair of transparent wings on mesothorax. Each wing is strengthened and controlled by a hooked appendage from the metathorax. The mouth or apparatus for taking food is wanting in the adult males, which are usually, instead, furnished with supplementary eyes.

Products of the Coccidae. While this family of plant parasites are hostile to the welfare of plant-life, they nevertheless give off some products of mercantile worth.

Several species of *Kermes* formerly afforded a red dye used by the Greeks and Romans. These insects dwell upon *Quercus coccifera* in the Mediterranean region. A medicinal syrup is also obtained from them. Several Coccidæ produce waxy matters. *Ceroplastes ceriferus,* one of the Lecaniinæ of India, produces white wax. The white wax of China is said to be produced by another Lecaniid, *Ericerus pela.* Little is known concerning this insect. It is understood that the wax is produced by the winged males. This wax was formerly much prized in China, but since the introduction of kerosene is falling into disuse. Lac is produced by *Carteria lacea,* still another Lecaniid in India, living on *Anona, Ficus, Rhamnus,* and other hosts. This lac is the shelly scale which the insect produces for covering. It is composed chiefly of resinous matter, with which there is mixed small quantity of wax, and some other substances. This insect's body yields the red substance called lake.

The European scale insects used for dyes were replaced after the discovery of America by the cochineal insects: *Coccus cacti*, of Mexico. This insect feeds on a cactus called Nopal (*Opuntia coccinellifera*). Later this insect was introduced into the Old World, and was established in a few places along the Mediterranean sea. In the Canary Islands it thrived so well on several species of cactus as to become an object of extensive commerce. The body tissues of these insects contained the colors from which the dyes were made. Aniline dyes have largely displaced coccid dyes. Axin, an external medicinal applicant, used also as a varnish, comes from the Mexican Coccid. *Llaveia axinus*. This seems to be a substance containing a peculiar acid, axinic acid. The so-called ground pearls are the encysted pupæ of Coccids belonging to the genus *Margarodes*. These chitinous cysts, in the island of St. Vincent, are of sufficient size to be collected and strung for necklaces.

Coccus (now Gossyparia) *mannifera* is a scale dwelling upon *Tamarix* in the Mediterranean basin. This scale exudes a honey-dew resembling honey. The Arabs use this honey-like substance for food, and call it "Man." It is supposed that this honey-dew or "man" is the real manna of the Israelites.

Collection. The necessary materials for a collecting trip are: a staff with crook handle, cigar-box, package of envelopes, hand-lens, notebook, and strong sharp knife. The staff is used to draw down high overhanging limbs of trees. The cigar-box swung over the shoulder by a strap serves as a receptacle for envelopes inclosing specimens. The hand-lens is for preliminary examination of the specimens. The notebook is to record data upon the specimens contained in consecutively numbered envelopes. The numbers in the notebook should of course correspond to the envelope containing specimens.

The notes should include date of collection, food plant (and here great care should be exercised to secure accuracy; a handy pocket edition of Gray's Field Botany is an almost indispensable companion), relative numbers present upon the host plant, and color of the scales. When the scales are on the trunk of the tree, the bark holding the insects can be removed. In case of twigs it will be more convenient to cut off the twig. When the laboratory is reached it will be found convenient to split out the center wood of the larger twigs. These twigs bearing specimens are ultimately to be placed in cork-stoppered, flat-bottomed test tubes (cork-stoppered bottles will do). Adapt the length of the wood then to the length of the test tubes. The twigs, bark, or leaves bearing scales are now ready to be placed in small pasteboard boxes. An assortment of prescription boxes I find indispensable. Empty spool-boxes will serve the purpose. The infested bark is to be left here until dry. If inclosed in the tubes before the sap is thoroughly dried out they will mould and materially damage the specimen thereon. While scales can be found in this latitude at any season of the year, the fall and early winter months are the best on account of the absence of leaves, and the presence of such forms as the adult Lecaniums, which die and fall off or are removed by sleets in the early spring months. Spring collecting should not be neglected, however, for this is the time when the males and nymphs can be most readily observed. At all times guard against the unnecessary removal of scales from their natural position on the cuticle of the host plant. The unskilled eye is sometimes deceived. Fungi and scars on the bark made by insects for the deposition of eggs, are mistaken for scale insects. Scale insects, not including gall, can be readily separated from the cuticle of the host plant without leaving an abrasion of the cuticle.

Preparation for study. Under the various divisions, preparation of specimens for microscopical study will be discussed. Since the distinguishing characteristics rest chiefly with microscopical structure, great care and close attention should be given to technique.

Terms used in discussion. In addition to the glossary found at the back accompanying the index, a brief preliminary discussion of terms may tend toward a better understanding for the beginner.

Scale has reference to the covering of the insects belonging to the Diaspinæ. In removing an insect of this family from its host, it will be observed that a scale or covering comes off apart from the body of the insect, and that beneath the insect there is frequently a thin whitish layer. This is the under part of the puparium and is known as the ventral scale. The part covering the insect is designated as the dorsal scale. In this work the term scale is sometimes used unqualified. In such cases it refers to the dorsal scale. In the Diaspinæ the last abdominal segment, frequently called the pygidium, bears the stable distinguishing characteristics. The lobes, plates, spines, circumgenital and dorsal glands are illustrated in the accompanying diagrammatic figure (fig. *a*), showing their relative positions in the genus Chionaspis. Upon these rest the

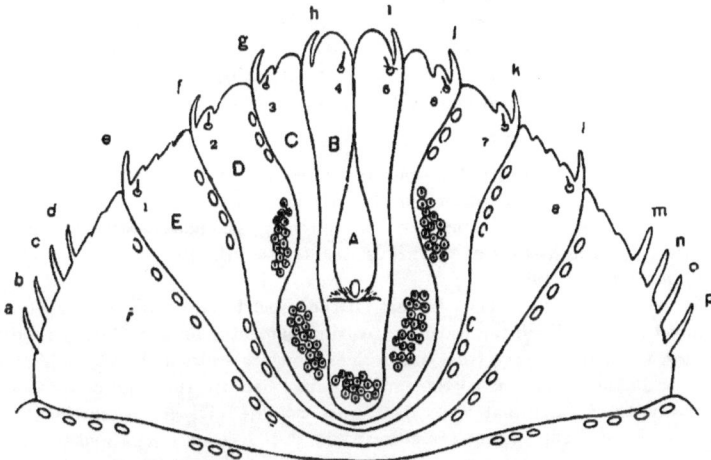

Fig. a.

Diagrammatic plan of pygidium of Diaspinæ, typical of the genus Chionaspis. *A*, median space, *B*, *C*, *D*, *E*, *F*, first, second, third, fourth, and basal spaces respectively. At the base of *A* is the circular anal opening; the transverse line just below is the genital opening. Grouped glands (spinnerets) in *B* are the median, in *C* the anterio-lateral, in *D* the posterio-lateral,—all on the ventral surface. *a*, *b*, *c*, *d*, *e*, etc., are simple plates; in many of the Diaspinæ these are toothed. 1, 2, 3, 4, 5, etc., are spines on the dorsal surface. The dorsal pores are arranged in rows in *D*, *E*, and *F*. The median lobes are at the caudal extremities of *B*, the notched or cleft second lobes are at the caudal extremities of *C*, the third pair of lobes are similarly located and notched in *D*.

characteristics which are to be most depended upon in the study of differentiation. The scale itself is an unstable character in the Diaspinæ. In the Lecaniinæ, as will be seen, the chitinous covering is more constant. The relative values in differentiation of these various structures are discussed under the species enumerated.

Classification. The separation into subfamilies of the family Coccidæ differs with different authorities. Signoret arranges the genera and species under four divisions. Green groups them under ten subfamilies, and Cockerell under eleven. It is not within the province of this article to enter into the merits of the various systems, but simply to refer to those subfamilies under which the species hereinafter discussed come. The only question which might arise would be the disposition of the forms placed under the Coccinæ, an unsatisfactory group at best as now arranged.

TABLE FOR DETERMINATION OF SUBFAMILIES.

A. Caudal portion of abdomen of female terminating in chitinized compound segment, the pygidium. Anal orifice without setiferous ring.
  1. Adult female without limbs. Insects possessing distinct scale covering composed of exuviæ and secretions.   Diaspinæ.
AA. Pygidium wanting in the female. Female insects naked, or covered with shields of waxy, chitinous or cottony secretions.
  B. Setiferous ring around anal orifice.
    1. Caudal extremity cleft; a pair of triangular hinged plates extends over anal orifice; triangular plates often setiferous. Limbs in adult female generally functionless. Body of female naked.   Lecaniinæ.
    2. Caudal extremity not cleft. Limbs functional throughout life. Body of female protected by cottony secretion, sometimes arranged in tufts.   Ortheziinæ.
  BB. Setiferous anal ring wanting.
    1. Abdomen of female terminating, in all stages, in prominent processes. Body of female shielded by cottony, or felted secretion.   Coccinæ.
    2. Abdomen of larval stage, only, terminating in marked prominences. Females naked or shielded by waxy secretion.   Hemicoccinæ.

DIASPINAE.

The subfamily, Diaspinæ, the armored scales, have a separate protecting scale, covering the body and composed partly of the exuviæ and partly of secreted matter. The female scale is approximately circular; the male scales are generally elongate. The life history of the genus *Aspidiotus* will illustrate in general the life cycle of this group.

The members of this genus in this latitude spend the winter as almost fully developed insects. With the warmth of spring, about the first of May, maturity is attained, the males emerge from under their elongate scales and seek their mates. The eggs, in the oviparous species, are deposited beneath the scale of the female, to remain there until hatched. The most notable viviparous species is the San José scale. The oblong larvæ have six well-developed legs and a pair of feelers or antennæ. The sexes, at this stage, are similar. The larvæ (Plate V, fig. 23) creep about in quest of suitable places to locate upon the host plant. They insert their long, filamentary sucking-tubes into the tissues of the plant. In the positions chosen the females remain for life, since they early shed their larval skin, and with the skins, the feet, antennæ, and eyes. Even before this a waxy secretion appears upon the back. At this first moult the larval skin splits around the lateral margins, separating the dorsal and ventral halves of the insect; with the ventral half go the sheaths of the antennæ and limbs. The insect grows rapidly, and covers its increased size by waxy filamentary secretion, added around the margin of the dorsal larval exuviæ. At the second and last moult the skin splits as before. This covering now becomes more chitinous, and still serves as protection for the insect. As the female increases somewhat in size after this time, the additional covering necessary is provided for again by marginal additions of the filamentary secretion from special dorsal organs for that purpose. The male scale is generally more elongate, and under it the male passes through complete metamorphosis. While the scale shows but one ix

larval pellicle, several skins are shed, the later ones being pushed out from beneath the scale. The male comes forth a delicate two-winged insect with six legs, without food-taking apparatus, and bearing two simple eyes in the usual position for the mouth.

The life-history of but few of these scales has been carefully worked out. This phase of the subject offers a fertile and profitable field for biological study.

Genera.

A. Scale of female circular, exuviæ central or approaching the margin.
    1. Scale of male oblong; color and texture similar to that of female; exuviæ centrally located, sometimes noticeably nearer one of the extremities.   Aspidiotus.
    2. Scale of male elongated; exuviæ at one extremity.
        (a) Male scale white and carinated.   Diaspis.
B. Scale of female elongated; exuviæ terminal.
    1. Scale of male white, shorter and narrower than female scale. In pygidium five groups of circumgenital glands present.   Chionaspis.
    2. Scale of male similar in form and texture to that of female.   Mytilaspis.
    3. Scale of female shows distinctly two moulted skins. Scale of male narrow, elongate; scale of female sometimes oval.   Parlatoria.

TABLE FOR DETERMINATION OF SPECIES.

Aspidiotus.

A. Female pygidium with three pairs of lobes.
B. Third lobe much reduced or only a pointed prominence.
  C. Size of scale 1½ to 2 mm.; scale somewhat convex.   fernaldi.
  CC. Size of scale 2½ mm.; scale flat.   juglans regiae.
B. Third lobe about same size as second lobe.
  C. Three pairs of incisions (see Plate VI. fig. 25), with distinct chitinous processes and 5 groups of spinnerets; plates simple and inconspicuous; exuviæ black under yellowish brown covering.   obscurus.
  CC. No incisions or chitinous processes; 4 groups of spinnerets; plates conspicuous and toothed; exuviæ covered with thin yellow film.
        hederae nerii.
AA. Female pygidium with less than 3 lobes.
B. Female pygidium with third pair of lobes wanting.
  C. Spinnerets present, median lobes converging at the tip; chitinous processes unequal, inner one larger (see Plate I, fig. 1); scales arranged in clusters; exuviæ of male scale yellowish when rubbed.   forbesi.
  CC. Spinnerets wanting; median lobes but slightly converging at the tip; chitinous processes equal (see Plate V, fig. 19); exuviæ of male scale black when rubbed.   perniciousus.
BB. Female pygidium with 2d and 3d pairs of lobes wanting or rudimentary.
C. Spinnerets present.
  D. Four groups of spinnerets.
    E. Scale light in color; exuviæ naked, pale straw color; plates toothed.
        greenii.
    EE. Scale dirty gray; exuviæ orange-colored, covered with gray; plates simple.   osborni.
  DD. Five groups of spinnerets.
    E. Exuviæ orange-colored; scale dark gray.

F. Scales scattered; size of scale, 1.5 to 3 mm. *aesculi.*
FF. Scales arranged in clusters, meeting on cephalic margins; size of scale, 1.4 x 1.3 mm. *ancylus.*
EE. Exuviæ yellow, covered with white; scales light yellowish brown. *uvae.*
CC. Spinnerets wanting; scale convex; exuviæ orange yellow, covered with white. *ulmi.*

Chionaspis.

A. Inner margin of median lobes widely separated.
B. Median lobes parallel, separated by one-third width of lobes. *pinifoliae.*
BB. Median lobes diverging from base.
C. Posterior group of 2d row of dorsal glands absent; minute circular glands absent (see Plate XV, fig. 1); male scale not carinated. *platani.*
CC. Posterior group of 2d row of dorsal glands present; minute circular glands present; male scale tricarinated. (Plate XIV, fig. 1.) *salicis nigrae.*
AA. Inner margins of median lobes close together.
B. Inner margins of median lobes fused; 2d row of dorsal glands absent; scale broadly convex. *americana.*
BB. Inner margins of median lobes separated, but almost contiguous.
C. 2d row of dorsal glands absent; scale flat or slightly convex; male scale tricarinate. *furfura.*
CC. Only anterior group of 2d row of dorsal plants present; scale quite convex; male scale not carinated. *ortholobis.*

LECANIINAE.

Under this subfamily there are two genera represented, Lecanium and Pulvinaria. In both the female is naked. In the former the eggs remain under the body of the female, while in the latter a cottony ovisae extends posteriorly from under the body of the scale. The larvæ of both sexes in this subfamily have a pair of large anal lobes. (See Plate XVI, fig. 3.) At this stage they possess well-developed legs, and are active. Later, when these young scales have settled down, these legs and the antennæ become functionless, and may be observed in the bleached body or test of the adult insect. Further details are brought out in the discussion of the characteristics of the various species.

Lecanium.

A. Scale with 2 prominent tubercles. *cockerelli.*
AA. Scale without tubercles.
B. Scale dark brown or black.
C. Antenna six-jointed. *kansasense.*
CC. Antenna 7- or 8-jointed.
D. Scale with ridges forming H, small ridges extending half-way up from margin; 2 tarsal digitules. *oleae.*
DD. Scale without ridges forming H, caudal part of body plicate; only one tarsal digitule. *canadense.*
BB. Scale light brown or yellowish brown.
C. Scale low or nearly flat.
D. Scale low, convex, pitted around fusiform median surface and on sides of scale, sides in folds and depressions; caudal margin plicate. *aurantiacum.*

DD. Scale nearly flat, few punctures, pits or folds; caudal margin not
plicate. *hesperidum.*
CC. Scale highly convex.
D. Scale measurement: length 2½ to 3⅞, lat. 1¾ to 2⅟₃, alt. 1 to 1¾;
found on tropical food plants. *coffeae.*
DD. Scale measurement: length 4, lat. 3, alt. 2.5 mm.; native of tem-
perate zone. *armeniacum.*

LIST OF SPECIES.

Diaspinae.

Aspidiotus forbesi.
ancylus.
uvæ.
Osborni.
ulmi.
fernaldi-albiventer.
obscurus.
juglans-regiæ.
perniciosus.
greenii.
hederæ-nerii.
æsculi-solus.
Mytilaspis pomorum.
Diaspis snowii.
Chionaspis ortholobis.
salicis-nigræ.
americana.
platani.
pinifoliæ.
furfura.
Parlatoria pergandei.

Lecaniinae.

Lecanium aurantiacum.
canadense.
kansasense.
cockerelli.
armeniacum.
hesperidum.
coffeæ.
oleæ.
Pulvinaria innumerabilis.
pruni.

Coccinae.

Lecaniodiaspis parrotti.
celtidis-pruinosus.

Hemicoccinae.

Kermes pubescens.
rivalis.

Ortheziinae.

Orthezia graminis.

The Coccidæ of Kansas.

Contributions from the Entomological Laboratory. No. 64.

Author's Edition, published December 17, 1898.

BY S. J. HUNTER.

With Plates I to VII.

The Homopteran family, Coccidæ, in many respects anomalous and in others similar, is unique in the diversities of the sexes and the modes of distribution. Sex and genus necessarily present differences. The variations existing within a single species, and gradations between species, however, have proven of peculiar interest. It is with these specific differences in view that the studies upon this group have been prosecuted. It is believed that from a large series of discriminations interesting deductions might be made.

The manner of preparation and method of study of this group have occupied some time. I will endeavor to give briefly an outline of my observations and experiments upon the technique of the genera mentioned in this paper. The first work was done with the usual bleaching reagent, a strong solution of potassium hydrate, washed in water, transferred to fifty per cent alcohol, then to ninety-five per cent alcohol and from here to a clearing mixture composed of two parts by measure of carbolic acid crystals and three parts of rectified oil of turpentine, taken from this to slide and mounted in xylol-balsam. The difficulties found in this process were: dangers of boiling too much or of not boiling enough, specimens had to be removed while the liquid was warm else a deposit would collect upon them and render them useless, material was generally more or less macerated and frequently the plates were not retained throughout the process.

Another means used was to soften the specimens in warm water then transfer to fifty per cent alcohol, then to ninety-five per cent alcohol, from here through the clearing mixture into balsam. This I found better, in that it insured the retention of the plates present but it had the objection of not rendering old and heavily chitinized scales transparent.

For a new bleaching agent, chloral hydrate, saturated solution, boiling specimens in this under cover for a few moments and also by leaving in the fluid under cover for twelve to twenty-four hours. The latter method —leaving in for a day or less—gave the best results, rendered the specimens transparent and kept them reasonably firm. It did not appear to affect the plates and spines; did not leave any undesirable substances upon the specimens. From the chloral hydrate they were taken into the water and then through the process previously described.

In experimenting with the action of xylol upon the plates it was found that it rendered them very brittle and left them undesirably transparent, so that in some specimens mounted in xylol-balsam I noticed a tendency to render the plates of the same refractive index as the balsam. Glycerine as a mounting medium was tried, and it was found that specimens could be more readily taken from a chloral solution into glycerine than by any other method used. The objection to the glycerine is, of course, that permanent mounts require to be incased in cement. Glycerine jelly, I think, will largely overcome this difficulty, and I expect to test it in my subsequent work.

Having the mounts ready for study I found daylight not always the most satisfactory light for revealing the points desired, so that the light of the Welsbach burner passed through a liter balloon flask, filled with a solution of copper sulphate, rendered the desired clearness by the addition of ammonia (NH_3), the test being a white light thrown upon the reflector. This illuminating apparatus is spoken of in *Botanisches Practicum*, Strasburger, 1897, and *Microscope and Microscopical Accessories*, Zeiss, No. 30, 1895. Mr. McClung, of the department of Zoology, in his work tried the addition of a drop or two of saturated alcoholic solution of saffron to the fluid in the flask. This I consider advantageous; enough being added to give the light a tinge of pink, which gives good contrasts with the yellow subject. I had endeavored to find a stain that would hold in firmly chitinized specimens, but met with little success. This pink light, showing the necessary contrast, as it does, served the same purpose at a great saving of time and labor.

The microscope used is the C. Zeiss, Jena, stand 1a, with complete equipment. The objectives used on triple nose piece are low power B, high power F, and oil immersion 1-12 inch, N. A. 1.22. Though I have tried an Abbe camera lucida, I prefer to make the sketches free hand.

Aspidiotus forbesi Johns. Plate I, Figs. 1, 2, 3, 4, 5: Plate II, Figs. 6, 7.

The scales of male and female in this group, brought together from widely separated localities, conform to the description.

Lots A, B, C, F and E show distinct tendency to congregate in groups and then others mass upon these. Lot G, though on old cherry trees, were few in number and only to be found upon one or two spots in three adjacent cherry trees. Lot F was most abundant upon the old bark of the trunk of the apple tree. Bark of apple tree frequently pitted. Lots D and E appeared to be distributed more nearly over the bark. Of this group the following numbers of mounts of each have been studied: eight, of A; ten, of B; nine, of C; four, of G; ten, of D; seven, of E; four, of F; making in all fifty-two individuals, these agree with "Plates inconspicuous or absent, spines prominent," but differ uniformly in number of spines.

There is in each lot, spine on median lobe agreeing with description, *but another spine appears extending over the margin from the dorsal side*, one beside the "second" and one beside the "third" spine, making a pair of spines, one on the ventral and one on the dorsal side, instead of one spine at "second" and "third." This was so in all cases and in this respect they agree with *ancyclus*. The chitine at the incisions appears to be uniformly heavier than shown in the original sketch. Lot D presents the greatest variations, as shown by sketches. Another form, which is very suggestive of another species, appears on same branch among this lot D. More material will be required before this point can be satisfactorily determined. The second lobe is notched, agreeing with Professor Cockerell's observation, but the larger chitinous process between first and second lobe is less twisted than shown by the same author. (Tech. Ser. No. 6, Div. Ent., pp. 6 and 8.) The scale is found upon trunk and branches. More abundant upon heavier limbs. I have found it upon the twigs and in and around the terminal buds of twigs.*

*Since the above was written I have found this scale upon cherry and apple in Anderson county and upon cherry and apple in Franklin county. In Franklin county a plot of five acres of apple stock being cultivated for scions, was so generally infested by this scale that it was deemed advisable to uproot the whole plant. A large cherry tree near by was also condemned. The scions cut this fall from the apple plot will be subjected to gas treatment before being used.

Forbesi was first reported from Kansas by Prof. E. A. Popenoe in *Kansas Farmer*, locality not given, reference to which is made in my Bull. Dept. Ent. Univ. Kans., *Scale Insects Injurious to Orchards*, p. 24, Jan., 1898.

Lot A. on cherry, Lawrence, common, taken January 28, 1898.

Lot B, on cherry, Seward county, taken February 18, 1898.

Lot C, on cherry, Cloud county, taken March 12, 1898.   (No sketch.)

Lot G, on cherry, 2½ miles northwest of Lawrence, taken October 18, 1898.

Lot D, crab apple, Lawrence, taken April 4, 1898,

Lots E and F, apple, Lawrence, taken April 5 and 20, 1898.

Aspidiotus ancyclus Putnam. Plate II, Figs. 8, 9.

Common on maple. In this species, as in the most of the *Aspidiotus* I have observed, there is a marked tendency to segregate. Clusters of scales sometimes two and three deep are to be found while comparatively few scales are situated alone. While this is doubtless true to a greater or less extent in all, it seems to be more noticeable in this species. And further that the scales meet upon the cephalic margins. Number of specimens of 3b studied, 12. The contour of anal plate of 3a and 3b agree. Second lobe depressed. Differences and locations of spines and plates are shown in the accompanying drawings.

The circumgenitals in 3b exceed the limits given for the species. Caudo-laterals 7 to 10, cephalo-laterals 13 to 19, anterior 0 to 4. 3b differs from 3a further in the chitinization of the incisions, 3a agrees with original figure in having the chitine nearly evenly distributed over the incision, almost crescent shaped, 3b shows the chitine along the sides of the incisions and incisions clear at bottom. 3b shows more dorsal glands, and this was characteristic of this group.

Common on maple in and around Lawrence.

Aspidiotus uvæ Comst. Plate III, Fig. 10.

On grape vines. No marked variations apparent. Iola, Kansas, February 12, 1898.

In June Mr. P. J. Parrott sent the writer specimens on grape from Manhattan.

Aspidiotus uvæ Comst. var. Plate III, Fig. 11.

Scale of female convex, circular, reddish white, red largely due to cork cells present, scale resembles *uvæ* somewhat; *uvæ* arranges itself in rows; this insect has no special mode of arrangement. Exuviæ slightly laterad of center, covered with white secretion; when this is removed orange colored exuviæ appear. Scale 1½ mm. in diameter.

Scale of male. Elongate, broad and roundly convex anteriorly and flattened posteriorly. The outline of margin of scale is egg-shaped, broad end at anterior margin, exuviæ covered with white secretion, nipple and ring fairly distinct. When rubbed rather large orange colored exuviæ appear. Exuviæ situated between the highest point and anterior margin, sometimes on anterior margin. Length 1 mm., width about ½ mm.

Female. Body circular, lemon yellow with many irregular deep orange spots, the greater number along the caudal margin of the penultimate segment. One pair of lobes notched on lateral margin, some specimens show lobes one entire the other notched. Approach each other at distal extremity, well chitinized, especially so along mesal margin near base. Two pairs of incisions with chitinized processes.

Plates forked, two caudad of each incision. One spine on median lobe, a pair laterad of each incision, and another pair about same distance laterad of second pair, as exists between first and second pair of spines, the third pair of spines are not as near to each other as the spines of the other pairs are. Five groups of ventral glands, numbers shown on figure 11. Nine mounts of females studied and many scales.

A further discussion upon this variety will be given subsequently. A comparison between a good series of the *Carya uvæ* with a series from *vitis* Sign. is desired. The exact status of this variety may then be better set forth.

On *Carya alba* Nutt. Lawrence.

Aspidiotus osborni Newell and Ckll. Plate III, Figs. 12, 13.

Since original description is not at hand for comparison I will add the following notes:

Scales rather evenly distributed over the branches of the tree. Scales slightly raised in center, depressed around margin. Color dirty gray, yellowish on top caused by exuviæ showing through, oval, 2 mm. by 1½ mm. Exuviæ laterad, covered by white secretion, when rubbed orange colored exuviæ appear; ventral scale white, very delicate.

Scale of male. Elongate oval, sides almost parallel. About 1½ mm. in length. Darker than female. Exuviæ laterad, lemon yellow when whitish secretion is removed. The scale is slightly raised along its greater median line, depressed at margin.

Female. Oval, lemon yellow. One pair lobes, median, showing slight notch on distal margin and faint notches may appear on lateral margin. Chitine appears to extend cephalad from base of lobes, mesal margins show chitinous processes.

Two pairs of incisions, each bearing chitinous processes on sides, processes nearly equal in size but rather wide apart.

Plates inconspicuous, one or two caudad of second incision, small. Spines prominent, one on lateral margin of median lobe, two between first and second incisions upon what might be considered a depressed lobe. One spine laterad of second incision and another spine half way between the fourth spine and a pair of spines on the lateral margin. Another spine appears on lateral margin near penultimate segment. (Not shown in figure.)

Four groups ventral glands, caudo-laterals 4 to 5, cephalo laterals 4 to 6.

From many scales and mounted specimens.

Nymph. Leg, antenna, and caudal margin shown by Fig. 13.

On *Quercus alba* L., Douglas Co.

Aspidiotus ulmi Johns.   Plate III, Fig. 14.

Found in two localities in the city of Lawrence upon catalpa, massed upon the branch of the tree. In an entirely different locality upon *Ulmus fulva*. It seemed in each case to be of long standing. Many old scales were found covered by the outer cork layer.

The scale of female agrees quite well with original description, except that I would not call the scale "quite convex;" slightly convex suits here better. Male scale agrees with description.

Mature female differs from original description and figure in having two notches upon the mesal margin of the lobes. In some specimens the proximal notch is faintly marked. The drawing made from the type shows here a slight curve.

The thickening of body wall as shown in original description is quite characteristic of specimens examined. In an examination of twenty mounted specimens no circumgenital glands were found.

Aspidiotus fernaldi Ckll. subspecies **albiventer** subsp, nov.   Plate IV, Figs. 15, 16.

Abundant on the trunk from the ground up. When the tree is wet the exposed ventral scales stand out almost like fine flakes of snow.

In the orange exuviæ, and in the mode of congregating and shape of scale, this insect favors *ancylus*. The color of the scale is lighter and the structure of the female precludes the possibility of its being *ancylus*. The Putnam scale prefers the branches, this scale the trunk. I have been unable to find a species described in the literature that agrees with this scale, so I have offered the following description:

Scale of female, grayish white, strongly resembling bark of maple, in clusters, ventral scale well developed, the white substance annulated, scale circular, somewhat convex with the cephalic margin extended, beyond the circumference, exuviæ cephalad of center covered by white secretion, when this is removed large dark orange red exuviæ appear. Scale 1½ to 2 mm.

Scale of male elongate oval, about ½ mm. in length, ventral scale well·developed, but not arranged in rings, color of dorsal scale somewhat darker. Exuviæ between center and cephalic margin, covered by white secretion, showing faint trace of dot and ring, when this removed orange red exuviæ appear. These exuviæ are much smaller than exuviæ of female scale.

Adult female, obovate, dark orange in color. There are three pairs of lobes. The median lobes are prominent, notched on lateral margin, line of mesal margin extends parallel with line of meson. Chitinous processes at inner base of lobes.

Second lobe pointed, erect, outer margin ranges from smooth undulating line to a margin bearing three distinct notches. This variation is shown by series of sketches (Fig. 16, a, b, c, d,). The third lobe is also erect, ranging from inconspicuous to long, slender, round lobe with notch (sketches Fig. 16, a, b, c, d,).

There are two pairs of incisions, with chitinous processes; these chitinous processes remain constant. A large club-shaped process on median side of first incision, a smaller one just opposite. The process in second incision approaches a crescent.

Plates are present, one or two inconspicuous caudad of first incision, always simple, one or two caudad second incision, generally forked. There is generally one simple plate between median lobes.

Spines, one on first lobe, usually two laterad of second lobe, the same may be said of the two situated laterad of third lobe. Another pair of spines are always present on the lateral margin between the third lobe and the penultimate segment.

There are five groups of ventral glands; caudo-laterals range from 5 to 8, cephalo-laterals 10 to 12, anterior 2 to 6. From many scales and mounted specimens.

The trunk of the maple bearing these scales was densely populated. Lawrence, Kans.

Aspidiotus obscurus Comst. Plate VI, Fig. 25.

On black oak *Quercus tinctoria?* Bartram. Douglas Co., Kans.

Aspidiotus juglans-regiæ Comst. var.   Plate IV, Fig. 17.

Scale of male, oval, elongate, 1 mm. in length, raised anteriorly, flattened posteriorly, darker than female, exuviæ situated near anterior margin, covered with white secretion, ring and dot within distinctly seen upon exuviæ.

Scale of female, diameter less than 3 mm., not more than 2½ mm. Ventral scale "mere film," "adheres to bark." Color and position of exuviæ agree fairly well.

Female.   Color pale yellow with "irregular orange colored spots." Oral setæ and last segment dark yellow. Number of circumgenitals do not agree with description, as shown by my drawing.

Lobes three, three pair present, pointed, plates "simple, inconspicuous," "resemble spines," the largest found caudad of each incision. Some specimens show simple plate between median lobes.

Incisions shown in figure. Mesal bases of median lobes slightly chitinized. Spines agree, except that on median lobe I find a spine not shown or mentioned by Comstock

Spines. First spine on mesal lobe, a pair between first and second incision, a third pair laterad of third lobe. This pair is also about half way between the median lobe and fourth pair of spines.

Plates inconspicuous, arranged as shown in drawing, one between median lobes, one cephalad of first incision, two cephalad of second incision, all simple.

Circumgenital glands, five groups, variable, extremes found are caudo-laterals 2 to 12, cephalo-laterals 8 to 13, anterior 0 to 5. Chitinized clubs present along side glands as shown. From ten prepared specimens of females and many scales of male and female.

On one young crab apple tree.   Lawrence.

A comparison between this species and the original description:

Juglans-regiæ var.

FEMALE.

Lobes two, and also a third pointed prominence, not to be called a lobe.

Median lobes vary in outline, some being more elongated than others. Well developed.

Second lobe about one-half as large as median lobe.

Second lobe has three notches on lateral margin, in all cases.

Two pairs of incisions, small but made conspicuous by chitinization.

A chain-like incision extends cephalad from first incision.

Plates simple and inconspicuous, resembling spines in form. Larger ones are one caudad of each incision.

Spines prominent.

Short spine on median lobe.

Two spines laterad of second lobe, both on ventral side.

Two spines laterad of third prominence, both on ventral side.

Pair of spines one-third of distance from median lobe to penultimate segment. Two spines on ventral surface near lateral margin of penultimate, and two on ante-penultimate segment.

Anterior glands 3 to 5, anterior laterals 9 to 11, posterior laterals 9 to 10.

MALE.

Antennæ, nine jointed including basal joint. Wing expanse about 2 mm.

Juglans-regiæ, as described by Professor Comstock, on p. 300, plate XIV, Agr. Rep. of 1880.

FEMALE.

Two lobes and a third pointed prominence.

Well developed median lobe, varying in outline.

Second lobe less than one-half as large as median lobe.

Second lobe with one or two notches on lateral margin.

Two pairs of incisions, small but made conspicuous by chitinization.

No mention.

Plates simple and inconspicuous, resembling spines in form. Larger ones are one caudad of each incision.

Spines prominent.

No mention.

Description does not say how many; drawing shows one on dorsal and one on ventral side.

Drawing shows a pair; one spine on dorsal and one on ventral side.

Fourth spines are nearer median lobe than to the penultimate segment.

Groups 4 or 5 in number. Anterior consisting of from 1 to 4, anterior laterals of from 7 to 16, posterior laterals of from 4 to 8.

MALE.

Not described.

Aspidiotus perniciosus Comst.   Plate IV, Fig. 18; Plate V, Figs 19, 20, 21, 22, 23.

The variations in the anal plate of female are greater than in any of the specimens of any species studied. In fact it was the exception to find one specimen having the structure of the two sides identical. The relative size of the chitinous processes between the first and second lobe remained the same, "close together and of nearly equal size," but these rarely ever appeared under the microscope at same local point—that is, the relative sizes of the two could be gained only by focusing up and down. Dorsal glands uniformly few in number, small plate just laterad of second incision always present.

A large number of individual mounts have been studied and sketches of many made to illustrate their structures.

The scales of this group are uniformly dark, in many cases black, the white secretion covering the exuviæ of male is scant, in some cases almost wanting, leaving the black sculptured ring and dot, resembling description of *A. andromelas*. Others, however, have the dot and ring fairly well marked.

The anal plate of female has presented many interesting features. Fig. I may be considered as an extreme. Very few indeed show the forks in plates so marked as they appear at a. This figure farther shows two plates caudad first incision, serrate, in this the specimen is normal. The irregularly shaped plates present on left side at b, between spines four and five, and absent entirely on right has been frequently observed. This figure further shows the inconspicuous plates between the median lobes. Their absence in some specimens examined could be satisfactorily explained by their being broken in course of preparation.

Fig. II may be taken as an average, shows only one plate caudad of first incision and this plate forked.

Laterad of spine four is another spine; this is unusual. Between spines four and five (not counting incidental spine) are two irregular plates on right and three on left.

Fig. III represents a left side of one specimen and a right side of another transposed and placed beneath for illustration of variations in structure of the two sides.

Figs. IV and V represent other individuals studied.

Newly born nymph. This agrees with description given by Howard and Marlatt in every respect save one. Their description reads: "The large central plates each terminate in a long hair." In the twenty nymphs examined the long hairs are not connected with the plates, but arise from the ventral surface of the body cephalad of the plates and extend out under the plates caudad.

On a pear tree brought from New Jersey. Location near Argentine, Kans.*

Aspidiotus greenii Ckll ‡ Plate VI, Fig. 24.

In the old scales found at base of palm leaf the exuviæ are nearly black, scale grayish black.

The plates found between the median lobes are in some cases forked. There is also a variation of two in the number of toothed plates. In some there are two more upon a side than in others. The greatest number of toothed plates found on a side was six, simple plates constant at four. Their relative position as well as the number and position of the spines is best shown by the figure.

Fifteen mounted females studied, and many scales. Found massed at base of palm leaves and extending a short distance up the leaf.

On the palm, *Howea belmoreana*, in green house. Lawrence, Kansas.

Aspidiotus hederæ Vall var. **nerii** Bouche.‡ Plate VI, Figs. 26, 27; Plate VII, Figs. 28, 29.

Scale of female and male agree with description. Position and condition of first and second skin of female as described.

Female, light yellow in color mottled with yellow.

The anterior and posterior lateral glands agree with numbers given, but in two specimens out of the twenty-five studied I find in the one case a pair of glands forming an *anterior group*. In another specimen I find an *anterior group* of two with one lying between this group and the right anterior lateral group. This I have shown in the sketch. This group of glands, not spoken of by Bouche or Comstock, is of interest. The plates and spines agree satisfactorily with the description. The comparatively small number bearing this anterior group may be the reason why this group, probably observed,† is not mentioned in the original descriptions.

*This tree was bought three years ago, when two years old, from Parry's Pomona Nursery, Parry, N. J. It was placed at northwest corner of a young orchard. It has been rooted up and burned. The writer visited this orchard on Dec. 10, 1898, and gave the grounds a careful examination; two small colonies of about a dozen individuals each were found on two trees adjacent, to place where infested tree stood. These have been cut away, and the owner, a progressive horticulturist, intends subjecting all the trees in that part of the orchard to a thorough treatment with whale oil soap wash as an extra precaution.

†I was unable to find reference in literature to median group in *Aspidiotus* s. str. I wrote Prof. Cockerell concerning my observation and asked for reference to such a group. He finds that Berlese examining *Aspidiotus hederæ* var *limonii* in Italy found one example with one gland representing the median group. Cockerell adds interestingly: "The appearance of a median gland thus is probably an atavistic feature, as the supposed ancestors of *aspidiotus* s. str. (*Diaspis*-like types) probably had the five groups."

‡As may be inferred from the host, these species are not indigenous.

Found massed on oleander leaves, much more abundant on under side than upper side. Atchison, Kans.

Aspidiotus æsculi Johns. sub. sp. **solus** sub. sp. nov. Plate VII, Fig. 30.

Female scale circular, 2½ to 3 mm. in diameter, flat, dark in central part, dirty gray around margin.   Texture of scale light; one side will adhere to the bark while the other side being loose will curl up over, scale translucent.   Exuviæ laterad of the center, sometimes on margin, covered with light secretion, easily rubbed off showing light orange exuviæ.   Ventral scale almost invisible to the naked eye.

The plates of the female agree in the main with *ancylus*, the contour of the anal plate resembles the same species, notably the depressed second lobes.   The scale presents differences which preclude a consideration of *ancylus*.   The scales are much scattered and are situated singly.   Upon seven twigs averaging three inches in length, all chosen from the tree because they bore scales, the following numbers of scales appear on each in the order examined: four, two, three, five, one, one, two.   On the twig bearing three, two of them were about one-fourth inch apart, the others were more remote from each other.

The male scale favors *forbesi* in color, but the description of *juglans-regiæ* in shape.

Mature female.   Light yellow, one pair of median lobes, notched midway of lateral margin, lobes in some converging, in others the inner line of lobe is straight or slightly divergent.   Two pairs of chitinous processes on each side, club-shaped.   From the pair just laterad of the median lobes there extend cephalad U-shaped openings in the body wall.

Plates prominent.   Caudad of first pair of club-shaped processes is a plate generally forked, between first and second pair processes there are from none to three plates, caudad of second pair processes are one or two plates.

There is a spine near lateral margin of lobe, a pair of spines between the two pairs of processes, a pair of spines just laterad of second pair of processes, another pair one-third distance from lobe to penultimate segment, a single spine a like distance from the penultimate segment.

Five groups of circumgenital glands, caudo-laterals 6 to 9, cephalo-laterals 9 to 14, anterior o to 2.   Club organs around glands, in some cases obscuring the anterior group.

From ten satisfactory mounts and many scales.   On *Juglans nigra* L.   University campus. .

Through the kindness of Professor Cockerell I have received some of the original material from which *æsculi* was described. I have compared the two species as follows:

A. sub. sp. **solus** compared with	A. **æsculi**, taken from the original material.
Scales of *solus* uniformly darker	than scales of *æsculi*.
Color of exuviæ of *solus* lighter	than exuviæ of *æsculi*.
Texture of scale very delicate.	Texture firm.
Scale flat.	Scale slightly convex.
Margin of scale irregular and frequently indistinct.	Margin of scale distinct.
Broad median lobe notched once on lateral margin; lobe bears spine.	Same.
Laterad of lobe is an incision bordered by club-shaped thickenings.	Laterad of lobe is an incision filled with chitine forming a crescent-shaped structure.
Line of U-shaped openings extending cephalad in body wall from the incision.	Same.
Caudad of incision one forked plate.	Same.
Laterad of this incision from two to five plates.	Laterad of this incision never more than two plates.
Laterad of median lobe on last segment are three pairs of spines, each pair composed of one spine from each side of the body wall, then a single spine near penultimate segment, making in all seven spines on lateral margin of the segment.	Only three spines on lateral margin of last segment and all arising from ventral surface.
Laterad of first incision is a second incision but little smaller than the first; the sides of this bear club-shaped chitinous processes.	The body wall at this point is entire and no additional chitine is apparent.
Dorsal glands numerous.	Dorsal glands numerous.

Five groups circumgenital glands. Caudo-laterals 6 to 9, cephalo-laterals 9 to 14, anterior 0 to 2. (Majority of specimens show 1 or 2 glands in anterior position.)

Four groups circumgenital glands. Caudo-laterals 8 to 11, cephalo-laterals 9 to 14, anterior group wanting.

The fifth group of glands, greater number of plates and spines, and additional well chitinized incision on margin of anal plate, together with peculiarities of the scale, are differentiations worthy of consideration in forming a specific separation. I prefer for the present, however, to place *solus* as a sub species, and have allied it to *æsculi* on account of similarity in contour of margin of anal plate of female. Later I hope to be able, with *A. juglandis*, *A. juglans-regiæ*, var. *pruni*, var. *albus*, var. *kafkæ* and *A. æsculi* before me, to establish the exact position of *solus*.

Mytilaspis pomorum Bouche.

The scales longer than measurements given in description, being 2½ to 3 mm. Structure of female conforms with typical forms. Found upon the outer branches of two apple trees in one locality. Lawrence, Kans.

Diaspis snowii, nov. sp. Plate VII, Figs. 31, 32, 33.

Scale of female. Flat, oval, 2 by 3 mm., exuviæ laterad of center on the shorter axis. Scale sometimes more nearly circular and more regular in outline, then exuviæ appear to be located nearer margin. Margins ragged. Central portion of scale dark gray, pallid on border, scale very flat approaching margin so that the connection between scale and bark is not readily discerned. A very delicate ventral scale.

Scale of male. White in fresh specimens, orange white in older specimens, 1 mm. long, narrow, rounded longitudinally, but slightly carinated. Exuviæ orange yellow, terminal. Some specimens not quite so long and broader. The mature male has not yet been observed.

Female. Oval, lemon yellow. Anal plate extended but little beyond the general shape, orange, well chitinized. Anal orifice situated at base of aperture between median lobes.

Circumgenitals, five groups, caudo-laterals 5 to 9, cephalo-laterals 8 to 15, median 0 to 2.

One pair lobes, notched on outer margin, two incisions on each side showing chitinous processes.

There is a spine on or near the lateral margin of the lobe, one between the first and second incisions, one laterad the second incision; the fourth spine is as far from the third as the third is from the median lobe. There is a small spine at a distance from the penultimate segment equal to distance between the second and third spines. On ventral margin of penultimate, and on each of the two segments cephalad, there are a pair of small spines. The rest of the body bears a spine on the ventral margin at somewhat irregular intervals.

Plates prominent, not quite as long as the spine, two caudad first incision, one or both forked, one forked plate caudad second incision and two plates, generally simple, between the incisions. Several rudimentary plates laterad second incision.

From eleven satisfactory mounts and many scales of both sexes. On *Salix nigra* Marshall. Douglas Co.

This species belongs to the subgenus *Epidiaspis* Ckll. MS., the type of which is *D. piricola* Del Guercio. The number of glands in the median group of *snowii* are less than in *piricola*, The glands of all groups are much less in number than in *D. pyri* Colvée. A distinction of moment is the extreme posterior position of anal orifice as in *Diaspidiotus*. This orifice is well cephalad upon the last segment in *piricola*, being located between the caudo-lateral groups of glands. The median lobes of this species show distinct notch about midway upon lateral margin, *piricola* is entire.*

It is highly fitting that this interesting representative of a European group, the first Coccid to be described from this laboratory, be dedicated to Chancellor F. H. Snow, whose persistent and untiring labors in the field of Natural History are largely responsible for the present status of biology in the University of Kansas.

My sincere gratitude is due Professor Theo. D. A. Cockerell for material encouragement and valuable suggestions kindly offered in the pursuit of these studies. I wish to express my appreciation of the assistance of Mr. P. A. Glenn, a student of this department last year, in the acquisition of the material here studied, and to acknowledge the earnest and careful work of Miss Ella Weeks in her part of the delineations accompanying this article.

The genus *Lecanium* is now under consideration and it is expected that a discussion of the Kansas forms will appear in a later number of the KANSAS UNIVERSITY QUARTERLY.

---

*For comparison between *D. piricola* and *D. pyri*, see Bull. Div. Ent. Tec. Ser., No. 6. p. 4, Cockerell.

The Coccidæ of Kansas, II.

Contribution from the Entomological Laboratory, No. 66.

BY S. J. HUNTER.

VIII to XII
With Plates ███ to ███.

In the study of the material here presented, it was found that the most satisfactory mounts were made from specimens boiled in a solution of KOH composed of equal parts of water and saturated solution of the potassium hydrate. The material was allowed to boil several minutes, then was transferred·to warm water and washed with camel's hair brush until all the coloring matter had left the body. From here the specimens were readily transferred to glycerine jelly for temporary study or through the clearing mixture mentioned in the previous paper into balsam for permanent reference.

I am again placed under obligations to Professor T. D. A. Cockerell who has very kindly read the manuscript and given some notes upon the species here studied. The drawings accompanying this paper were made by Miss Ella Weeks under the author's immediate supervision. Thanks are due her for the care and skill exercised in their production.

All measurements given, both in text and plates, are in micro-millimeters.

VIII

Lecanium maoluræ nov. sp. Plate ███ Figs. 1, 2.

Female. Scale; long. 3 to 5, lat. 2½ to 4, alt. 1½ to 2, μ. Color light brown, older scales comparatively flat, younger scales when dry wrinkle up forming ridges on longitudinal median line. In older specimens the longitudinal median surface is smooth, this area being fusiform, but not greatly enlarged in the center. Fusiform space pitted on each side, the lateral surface in folds; the depressions each contain several small pits. The caudal margin somewhat plicate. The edges of body wall not upturned on median margins of caudal opening. Gland pits in derm comparatively few. Mouth parts prominent, and bearing a well developed triangular labium having three prominent spines on lateral margins.

Antennæ seven jointed, the fourth being the largest, the sixth the shortest. Beginning with the proximal segment they measure 56-68; 32-36; 56-68; 52-56 28-40; 20-24; 28-32; micromillimeters respectively. Chætotaxy and structure shown in figure. In some specimens sutures between ɪ and 2, in others sutures between 3 and 4, 5 and 6 are indistinct. The third and fourth are without spines. the remaining segments show spines as indicated in figure.

The legs are well developed, and highly chitinized. The body when boiled in KOH gives reddish brown coloring matter and becomes clear, the legs however retain part of their chitine. Trochanter prominent, prothoracic leg in some specimens shows unusually long hairs, two on trochanter, one on coxa; hairs elsewhere as indicated in figure. Claw curved, with two stout knobbed digitules. The second digitule of claw seen only in the mesothoracic leg. Tarsus bearing two long slender knobbed digitules. Chætotaxy and relative sizes of segments shown in figure.

Habitat.[1] On the twigs of osage orange, May 18, 1898. These specimens were received from Claflin, Barber county, through the Honorable F. D. Coburn Secretary of the State Board of Agriculture. The twigs were thickly covered with scales but so thoroughly were they parasitized by chalcids and attacked by the larvæ of a coccinellid that it was with difficulty that suitable material for study was obtained from material sent. These insects are widely separated from *L. maclurarum* Ckll. Description of *L. robiniarum* Douglas is not at hand. I note however that a probable variety has been found in New Mexico* on osage orange. The parasites left no eggs for comparison, so even this remote clue is wanting. Professor Cockerell says: "It isn't *robiniarum*. It is related, I think, to *L. fitchii.*"

IX

Lecanium canadense Ckll. Plate ███ Figs. ɪ, 2, 3.

Lecanium caryae v. canadense Ckll. Can. Ent., Vol. 27, 1895, p. 253.

Lecanium canadense Ckll. Can Ent., Vol. 30, p. 294.

This scale at first glance resembles scale of *L. kansasense* but upon close examination the smooth central boss is not so apparent, nor raised smooth ridge on each side of the boss. The legs and antennæ present still greater distinctions. Scale of female, long, 4 to 5½; lat. 4; alt. 3½ to 4 μ. Hemispheroidal, and caudal margin slightly extended. Some specimens show slight elevation on center of dorsum, others not raised but shiny on dorsum and with but

*Ckll. Can. Ent. '95. p. 257. Insect Life. Vol. VII. p. 200.

few pits. Caudal half of lateral portion of body plicate. Color very dark brown, pits nowhere deep, unpitted surface shiny. Derm, when bleached shows numerous gland pores.

Antennæ six jointed. Professor Cockerell adds: "This species occasionally shows 7 jointed antennæ." The third joint about ⅓ length of whole member, the sixth shows distinct compression which suggests a suture but none could be found in specimens examined. Individual lengths and comparative sizes of segments shown in figure.

Leg, size of segments variable. Trochanter prominent. Limits of forms of this species studied:

C. 120-140; tr. 68; fem. 180-204 (including tr); tib. 120-152: tars. 80-88; cl. 12-12.   •

Only one long slender tarsal digitule seen, knobbed: two short stout digitules belong to the claw.

Habitat on *Ulmus americana* University Campus, April, 1898 Collected by Mr. P. A. Glenn.

Lecanium kansasense, nov. sp. Plate⬛ Figs. 4, 5

Female, long. 3 to 5⅛, lat. 2½ to 3¼ , alt. remains constant at 2, μ. Color dark brown, very shiny, derm apparently thick, when boiled in KOH becomes translucent retaining some coloring. Derm checkered with numerous gland pits. Some scales, when taken, March, '98, were sparsely covered with white powder. When scale is removed distinct oval ring with anal indentation distinctly apparent upon bark of host.

On median surface of dorsum is a smooth space, oval-shaped and slightly raised, bordered on each side by row of punctures. Some specimens show slight smooth elevation and another row of punctures laterad of first rows, so that boss from long focus appears to be encircled by two rows of pits. Lateral and caudal aspects plicate, cephalic aspect minutely wrinkled.

Antennæ six jointed, structure, average length, and arrangement of hairs shown in figure. Third joint a little longer than 4+5+6. Segments always well marked. Terminal hair equal in length to proximal joint.

Leg strongly chitinized, does not bleach as readily as derm. Number and position of digitules, average length of segments and chætotaxy shown in figure.

Habitat. On *Cersis canadensis L.*, University Campus, January, 1898.

I am inclined to associate with this species scales of *Lecanium*, insufficient in number for thorough study, taken from *Juglans nigra*, a potted rose, and from *Ulmus fulva*. In none of these were either legs or antennæ found. The paucity of material, therefore, prevents a more positive opinion. From the scales, however, it is evident that they are closely related, if not identical. Of the scale on *Ulmus fulva* Professor Cockerell says: "I feel confident that the one on *Ulmus fulva* belongs here." If these can subsequently be proven identical, it will be of interest in showing the range of adaptability of this species.

Lecanium cockerelli nov. sp   Plate X   Figs. 1, 2, 3.

Scale of female. Average long. 8.5, lat. 5 to 6 (apparently governed by width of branch), alt. 3, μ. Scales of extreme length are long. 10.4, lat. 6.6, alt. 4.2 This striking scale will be easily recognized by its two prominent tubercles on the cephalic half of the body, situated laterad of the longitudinal median line. From these tubercles the body descends rapidly cephalo-ventrad to the bark of host, giving the cephalic aspect somewhat of the appearance of the upper part of the face of a bull dog. From the tubercles the body slopes gradually caudo-ventrad to the bark of the host. Derm closely pitted with shallow punctures, color very dark brown.

When removed from bark the margin of the insect leaves an elliptical ring, central portion of which is covered by a white powder. By transmitted light derm is shown to be closely perforated by minute gland pores.

Antennæ, stout, eight segments, arrangement of hairs and length of segment shown in figure. Leg stout, tarsus bears two long slender knobbed digitules, claw with two stouter digitules, length of claw and tarsus 128 micromillimeters.

Habitat. Taken February, 1898, on *Ulmus americana* in Lawrence, Kansas. Many of the outer branches of trees were closely studded upon the upper surface by these scales. The same trees were examined February of this year and but a single specimen was found. I cannot charge this disappearance to parasites exclusively. I would rather believe that the severe weather and the two very heavy sleets that covered the trees and remained upon them for several days were largely responsible for the clearance of old scales.

This attractive species is dedicated to Professor T. D. A. Cockerell whose studies have greatly enriched Coccidological literature.

On walnut, the same species was found, but the scale is uniformly smaller and the size of segments of leg and antennæ correspondingly less. Long. 6-7, lat. 4-5, alt. 3-4, μ. A comparison of figure 2 with figure 3 will show differences in size of the antennæ and legs of the walnut and elm scales.

Habitat on *Juglans nigra*, University Campus. April, '98.

From the relative sizes can it be said that elm is the better adapted host?

Professor Cockerell writes that he has received specimens of this scale from Mr. G. B. King and several other correspondents but the material was in unfit condition for description,

Lecanium armeniacum Craw.   Plate, Fig. 4.

Comparison of the infested twigs with photograph upon title page of bulletin No. 83 Cornell University by M. V. Slingerland, suggests a possibility of this being the same species. Later, however, in the description, Professor Cockerell is quoted as associating that insect with *juglandis*, a 7-jointed species. There being no descriptions, specimens *in situ*, or slide specimens at hand the following notes are offered.

Scale of female, crowded closely upon twigs of plum sometimes 2 and 3, one upon another. Color of scale pale brown; some of scales full and show no foldings, others show a row of pits on each side of longitudinal median line and sides plicate. Some scales have retained their form while others are much shriveled up. The shape of scales, hemispheroidal with sides somewhat depressed. Long. 4, lat. 3, alt. 2.5, μ.

Antennæ 6 and 7 jointed, the basal joint quite stout and globular, the first and second segments bear unusually long hairs, the third segment (when 7 jointed) bears one, the fourth again bears two unusually long hairs. Further chætotaxy and relative dimensions shown in figures.

Legs are slender and might be characterized by the marked constriction at the joints. There is an exception to this, however; in the case of the trochanter and femur. Here the segments are simply marked off by a straight line, no noticeable indentation being apparent on the margins. Tarsus bears two long slender knobbed digitules, claw bears two much shorter and stouter digitules.

Habitat, on plum twigs. This insect was found by the writer among the collections and bore the label "Scale bugs on plum, Kansas." Reference to the lot number showed that these scales were taken by Prof. V. L. Kellogg in 1891. The date of collection was not given but upon twigs there remained some leaves and blossoms so that the time of the year can be satisfactorily determined.

A comparison with descriptions of *L. rugosum* Sig.* shows not "rugose" but plicate, not "hills" but ridges and furrows upon the sides. The plum scales are light brown. Antennæ 6 and 7 jointed, never "eight"; in view of obscurity of segmentation, it is interesting to note that the plum scale agrees with *rugosum* in that 3 exceeds part distad of 3, if 3+4 are one segment as in plate, figure 4 *a*, but if they are as drawn in same plate, figure 4 *b*, joint 3 is less than all distad. Size uniformly smaller than *rugosum*, plum scale elongate, *rugosum* circular. Groove in anterior tarsus not shown: posterior tarsus not wider than tibia; chætotaxy not similar.

With the Queenston scale:†—"legs well developed," trochanter has "one" hair, coxa bears not "one" but two long and two short hairs. Length of femur, tibia and tarsus within possible bounds, but digitules of claw not "long" but short and stout extending but little beyond claw.

No further literature being at hand for comparative study I sent this scale with my notes to Professor Cockerell. He says, "This scale has much in common with *L. armeniacum*, yet seems not quite the same. I wish we knew the sub- adult (living) female and the newly hatched larva." Professor Cockerell kindly sent me specimens of *armeniacum*. A comparison shows, legs similar; antennæ agree with seven segmented specimens except that 5th joint in *armeniacum* is shorter than in plum insect.

The greatest difference appears to be in the scale itself. When bleaching, it colors the fluid a yellowish ochre; the plum scale gives off brownish coloring. Scale of *armeniacum* is not plicate and shows no longitudinal median raised smooth ridge: this insect, *armeniacum*, is more hemispheroidal with side quite full; plum scale more elongate and flattened.

Professor Cockerell suggested that Mr. Theo. Pergande be consulted since he has been working upon these fruit tree Lecaniums.

Accordingly I have received the following from Mr. Pergande through Dr. Howard:

"I have examined and compared the specimens sent with mounted and dry material of *Lecanium armeniacum* in our collection, and have come to the conclusion that the plum scale from Kansas

---

*Translated by Mrs. T. D. A. Cockerell. Can. Ent. Vol. 27. p. 59.
†T. D. A. Cockerell. Ibid, p. 60.

is identical with the above species. With regard to the difference in length of one or the other of the antennal joints, as noticed, I will say that it is simply individual variation; even in the same specimen the comparative length of either of the joints of both antennæ varies frequently more or less. There is generally also a more or less perceptible variation in size, color and shape in the same species, dependent, in a measure, on the food plant on which it may have established itself, and also on the locality. Old specimens, which have attained their full growth and have died a natural death, are generally darker, if prepared for the microscope, than younger individuals of the same stage and with all the pores of the germ much more distinct. As to the shape of the individual scales and their sculpturing, I find in our material of typical specimens of *Lecanium armeniacum* the same variations as those mentioned."

The limitation of variations within a species never fails to be of interest. With a view to setting these forth in this species the following data are given concerning the antennæ:

After examining the antennæ of 19 bleached insects from plum by means of a $\frac{1}{2}$ oil immersion objective, it was found that four of this number bore 7 jointed antennæ, two showed faint trace of suture between 3 and 4 (of the 7 jointed variety) and thirteen bore distinctly 6 segmented antennæ, 3 and 4 appearing as one and about equal in length to 3 + 4 of the 7 segmented antennæ. Measurements of antennæ of representative insects will show these variations in detail.

A brace is used to show that the two antennæ belong to the same insect. In this connection it is well to note the variations in number of segments within the individual as shown in the cases of *f* and *i*.

Lot No. of Specimen.	1st	2d	3d	4th	5th	6th	7th
a	44	44	56	48	20	20	40
b	44	40	48	52	20	20	40
c	48	44	100		24	28	32
d	48	40	100		24	24	40
e	48	44	92		16	24	44
	36	48	100		20	24	40
f	48	36	108		20	24	44
	52	40	56	56	24	24	40
g	48	44	96		20	24	46
	44	32	108		24	20	44
h	48	28	92		20	20	40
i	44	40	96		20	24	40
	48	40	52	44	24	20	40
j	46	40	92		24	20	44
	44	44	92		24	20	40
k	44	40	88		24	20	40
	48	44	100		24	20	48
l	44	40	52	44	..	..	..

Antennæ of *L. armeniacum*[*] on prune, from Healdsburg, California (Ehrhorn).

m	48	40	56	48	20	16	40
	36	40	48	46	16	20	40
n	48	40	56	52	20	24	40
	36	40	52	56	20	24	40

Lecanium hesperidum L. Plate XVI, Figs. 1, 2, 3.

Habitat. In conservatories on *Abutilon* sp. and *Citrus* sp. In green house on *Citrus* sp., *Hedera helix* and *Nerium oleander*.

This species was found on trunk, branches, leaves and fruit of the citrus trees. The scales upon the trunk are darker and more convex than those on branches. Those on branches and leaves incline to be yellowish, while those on the trunk are grayish. The outlines of both remain the same (Plate XVI, Fig 1) except when modified by mid rib or by bordering on branch.

The scales on ivy and oleander are uniformly darker than orange scales; some of them are amber colored. The marginal outline is ovate, cephalic half being the wider. These scales are also more convex than the orange scales, and show on dorsum in some cases the figure H, the transverse bars joining the marginal spikes.

Antennæ, legs and nymphs of all these scales agree in structure. The subject of tesselation naturally arises in this connection. Considerable attention has been paid to the bleached derm. The results obtained, however, have not been constant. With the aid of favorable reagents and suitable stains it is hoped that some definite data may be secured which will admit of something being said upon this point at a later date.

Lecanium coffeæ Walker. Plate ▇▇ Fig. 4

Habitat. On sword fern, *Pteris* sp., green house, Lawrence.

Lecanium oleæ Bernard. Plate ▇▇ Fig. 5.

These scales differ in number and relative sizes of segments of antennæ as compared with Comstock's description in U. S. Ag. Rep., 1880, p. 336.

[*]Since the above went to press I have received the following concerning *L. armeniarum* on prune from Mountain View, California (Ehrhorn): "Antennæ (1) 47, (2) 28, (3) 45, (4) 58-62, (5) 18-19, (6) 20, (7) 45. One antenna appeared to have only six segments, 6 measuring 42 micro-millimeters. The legs were not obtained in good condition but the coxa is 60, the femur with trochanter 145 micro-millimeters " Cockerell and Parrott, Industrialist, April, 1899, p. 233.
An examination of five of the Healdsburg prune *armeniacum* gave cephalic leg—coxa 88, tr. 62, fem. 104, tib. 100, tar. and cl 70; meta-thoracic leg—coxa 96-104, tr. 56, fem. 128, tib. 120, tar. and cl 84.

Habitat. on *Nerium oleander*, green house, Lawrence.

Professor Cockerell notes upon this species: "The typical antenna has 8 segments. These examples vary in the direction of *L. mirandum*, Ckll. and Parrott, ined., from Tlacotalpan, Mexico, and a study of them, together with other material, has, led me to be of the opinion that *mirandum* is, after all, only one of the forms of *olea*."

Lecaniodiaspis (?) **parrotti nov. sp.**   Plate XII. Figs. 4. 5.

Turtle shaped back resembles somewhat the carapace of *Chelydra serpentina*. 7 tubercles compose median carina, the second the longest, then graduated dorsad, radiators extend down and out from tubercles. Ribs on dorsum apparent, corresponding in number, position and size to the median tubercles. Ribs bear distinct elevations midway between carina and margin of body, and where ribs meet margin are to be seen protuberances corresponding in size to the median tubercles.

Marginal outline forms an oval; cephalic margin bears three small tubercles, the median pointed, the ones on each side obtuse, the three being nearly equal in size. At the caudal extremity of longitudinal median carina is a prominent quadrangular structure, extending caudad from margin of the body.

Thickness of body 1.1μ., margin of dorsum elevated from bark $.4 \mu$. Dorsum and side wine colored, covered in places by grayish white, derm of body wrinkled, waxy secretion apparent under ventrum.

Described from one specimen *in situ* on *Aesculus glabra*, and named in honor of Mr. P. J. Parrott of the State Agricultural College. Taken 4 miles west of Lawrence, February 9, 1899.

This specimen was opened and within there was found the pupa of a parasitic hymenoptera, which in its development had destroyed the larger part of the body of the insect, so that it could not be ascertained whether the antennæ were present, rudimentary or absent. It is therefore placed provisionally in *Lecaniodiaspis*. The segmentation and chitinization of ventrum of that part of the body examined resembled the same portion of the body of *pruinosus*. The scale itself is quite characteristic and will, I believe, be readily recognized from the two figures accompanying this description. In this connection it is fitting to record that Professor Cockerell wishes to say that in his opinion *Lecaniodiaspis artemesiæ* Ckll should be transferred to *Solenococcus* as *Solenococcus artemesiæ* Ckll.

Lecaniodiaspis celtidis Ckll., sub. sp. **pruinosus** sub. sp nov. Plate XVII, Figs. 1, 2.

Female: long. 4, lat. 3, alt. 1, μ Median carina and interrupted transverse ridges distinctly seen in younger specimens, less apparent on dorsum of old scales. Scales slightly convex, sides of dorsum arise from bark of host. Color pale ochrous with frosty white covering. The color and shape resemble *celtidis* but in *celtidis* the frosting is coarser and *celtidis* does not show median carina or segmentation on dorsum.

Anal margins bearing hairs, two bristles prominent on tips of each anal plate. Antennæ eight jointed, joints distinct, measurements and chætotaxy shown in figure.

Ventrum distinctly segmented showing chitinized plates transverse and longitudinal. Mouth parts prominent, setæ 1.48 mm. long.

Antennæ 8 jointed, basal joint, 36; 2d, 24; 3d, 36; 4th, 40; 5th, 40; 6th, 28; 7th, 24; 8th, 20.

Habitat on bark of *Ulmus americana*, Lawrence, April, 1898.

I have not seen the original description of *L. celtidis* Ckll. but I received some time ago specimens of *celtidis* from Prof. Cockerell, and have made therefrom a sketch of the antenna for comparison. Antenna of *pruinosus* is eight segmented; the basal segment bearing one minute hair, the second has two prominent hairs, third and fourth joints naked; fifth, six and seventh bearing each one hair; the terminal segment carries seven bristles. The whole member is short and stout. The eight jointed antenna of *celtidis* is longer, more slender, and bristles were observed on the second, fifth and terminal joints, the terminal joint bearing eight. The differences in length of respective segments are shown by the measurements attending. The characteristic structures of the scales, as before stated, are distinctions of moment.

The genera, *Chionaspis* and *Pulvinaria* will be discussed in the next paper upon this subject.

Author's edition, published April 12, 1899.

COCCIDÆ OF KANSAS, III.

Contribution from the Entomological Laboratory No. 72.

BY S. J. HUNTER.

With Plates XIII to XIX, inclusive.

Chionaspis ortholobis Comstock. Plate XIII, figs. 1, 2.

On willow (*Salix* sp.) on bank of Kaw river, and cottonwood (*Populus* sp.) in the vicinity of Lawrence; also on willow (*Salix*) on University campus.

Chionaspis salicis-nigræ Walsh. Plate XIV, fig. 1.

On willow (*Salix* sp.) near Lost Springs, Marion county, and on host of same genus near Greeley, Anderson county.

In my study of these two species I have found the differentiation (when the mature female only was represented in the material at hand) attended by some uncertainty. In looking over the literature I find the same difficulty sometimes expressed. The distinctive characteristics of each I have endeavored to set forth. The comparisons are based upon an abundance of material determined by the writer from the Kansas localities given, and upon the following from the Division of Entomology at Washington: Specimens of *C. ortholobis* from San Bernardino, Cal., the type locality, and *C. salicis-nigræ* from Mankato, Kan., authentic material, which through the courtesy of Mr. C. L. Marlatt I have had the privilege of studying in this connection.

C. ortholobis.	*C. salicis-nigræ.*
SCALE OF MALE.	
Without carinæ.	Tricarinate.
Exuviæ dark yellow or brownish.	Exuviæ pale lemon yellow or colorless.
SCALE OF FEMALE.	
White.	White.
1.8-2.2 mm. long.	2.6-3.4 mm. long.
Exuviæ yellowish brown.	Yellow or colorless.

PYGIDIUM OF FEMALE.

Median lobes almost fused in basal half; inner margins frequently divergent in distal half, serrate.

Second lobes half or less than half as long as median lobes.

Space between second and third lobes less than twice the distance between the first and second lobes.

Third lobe, inner lobule less than one-half inner lobule of second lobe.

Plates 1, 1-2, 1-2, 1-2, 1-4.

Circumgenital glands: Median, 17-24; post. lat., 14-27; ant. lat., 21-33.

In second row dorsal glands, posterior group absent.

Anteriors 2-6; minute circular glands absent.

| |
Median lobes short, broad, symmetrically rounded at tips.

Second lobes half or more than half as long as median lobes.

Space between second and third lobes more than twice distance between first and second lobes.

Third lobe frequently prominent; inner lobule = one-half inner lobule of second lobe.[1]

Plates 1, 1-2, 1-2, 1-2, 1-5.

Median, 23-30; ant. lat., 17-53; post. lat., 19-40.

Posteriors 1-4.

Anteriors 2-4; minute circular glands 1-10.

Chionaspis salicis-nigræ, readily separated from *C. ortholobis* by its male scale, can, from observations made upon the pygidia of fifty-four females, be said to differ in this structure by its shorter and broader lobes, by its possession of the posterior group of dorsal glands in the second row, and the presence of minute circular glands most abundant in the anterior group, second row, and visible beneath (ventral aspect) the circumgenital glands. In an examination of twenty-seven mounts of *ortholobis* and twenty-seven mounts of *salicis-nigræ*, I found this posterior group in second-row dorsal glands absent in all *ortholobis*, and present in all but one *salicis-nigræ*; this individual *salicis-nigræ* was well marked with the minute circular dorsal glands. The statement of Comstock[2] concerning median lobes of *ortholobis*, "mesal margins are parallel for more than half their length," holds good in many individuals among *salicis-nigræ*. "The distal margin of each (*ortholobis*) is rounded"[2] does not always obtain, since the inner distal margin is frequently divergent while the outer margin is rounded.

Chionaspis americana Johnson. Plate XIV, fig. 2.

Scale of female, 2 to 3.5 mm. in length: exuviæ 0.8 mm. long, sides diverging; generally straight, sometimes curved to right or left, curved scales located singly, broadly convex, dirty white.

Scale of male, tricarinate, 0.7 to 1 mm. in length, sides nearly parallel, clear white; exuviæ pale lemon yellow.

Mature female. The pygidium bears three pairs of lobes. Median lobes prominent, fused almost entire length of inner margins, sloping rapidly laterad, lateral margins with from one to three distinct

1. Cooley says "third pair often almost obsolete." Sp. Bull. Hatch Exp. Sta. Aug. 1899, p.
19. When this is the case lobes 1 and 2 are proportionately depressed.
2. Rept. U. S. Comm'r Agr. 1880, p. 317.

notches. Second lobes prominent, consisting of a large mesially inclined inner lobule, showing one or two marginal notches, and a shorter, more erect, outer lobule, sometimes with one notch. Third lobe generally compressed, though clearly distinguishable, division into two lobules apparent, inner lobule generally entire, margin of outer lobule notched once or twice, spines appear singly, as shown in figure 2, plate XIV, and grouped, beginning lateral of the median lobes, as follows: 1, 1-3, 1-2, 2-4, 4-8. The circumgenital glands range in number: Median groups, 16-26, anterio-laterals, 15-36, posterio-laterals, 15-39. Of the thirteen specimens in which the glands were counted, ten had decidedly more glands upon the right side than upon the left side. Location and number of the dorsal glands shown in the figure.

Male. The author of this species, Prof. W. G. Johnson, records two forms of males —"A perfect male with fully developed wings, and a pseudo-imago with rudimentary wings."[3]

This species is very common in Lawrence and vicinity, where I have taken it chiefly upon the outer branches of the white elm, but have also found it existing upon the trunk of the tree. I have received it from Floral, Cowley county, and Abilene, Dickinson county, upon badly infested twigs of the white elm.

In the abundant material studied, from five localities in Douglas county, and at Abilene and Floral, little striking variation is to be noted. The plates are sometimes forked and sometimes simple, the limits of circumgenital glands rather large; lobular crenulations appear at irregular intervals. The structural characteristics of the species in this latitude, however, appear to be fairly constant.

Chionaspis platani Cooley. Plate XV, fig. 1.

Scale of male, 0.9 to 1.3 mm. in length, sides parallel, diverging, slightly oval, without carinæ, exuviae lemon yellow, occupying about one-fourth the length of scale.

Scale of female, 1.4 to 2 mm. in length, broadens posteriorly sometimes abruptly, color favors the whitish coloring of the bark of host, sometimes obscured by the characteristic pruinose coloring of the bark; exuviae dark reddish brown, prominent, about one-fourth of length of scale; little or no ventral scale.

Female. Pygidium bears three pairs of lobes; median lobes prominent, divergent from inner base, plainly serrate; second lobes consist of two lobules, the inner lobe the more prominent, the outer lobe extending but little if any beyond the marginal outline, faintly serrate; the third pair obscure, inner lobule noticeable, the outer represented are arranged lateral of median lobes, beginning at the median lobe 1, generally as a bench-like lateral extension of the inner lobule. Plates 1, 1-2, 1-2, plates simple, spines and dorsal glands arranged as shown in figure, together with an average number of circumgenital glands.

Since this scale is described from Kansas, the specimens in hand conform with Cooley's description. The male scale seems to be uniformly larger than in the type insect, and with darker exuviæ.

Chionaspis pinifoliæ Fitch. Plate XV, figs. 1, 2.

Common upon *Pinus* sp. on the campus in Lawrence and vicinity. The infestation nowhere serious.

Pulvinaria innumerabilis Rathv. Plate XVI, figs. 1, 2.

On soft maple (*Acer* sp.), white elm (*Ulmus americana*), honey-locust (*Robinia* sp.), black walnut (*Juglans nigra*), in the vicinity of Lawrence, and on maple near Kansas City.

MEASUREMENTS IN MICROMILLIMETERS.

HOST.	ANTENNAL SEGMENTS.							
	1	2	3	4	5	6	7	8
Maple................	45	48	63	51	36	27	24	45
Honey-locust............	39	42	69	45	30	24	21	45
Walnut......	45	48	63	51	36	27	24	45
"	45	51	66	60	36	24	24	42

	LEGS.					
	Coxa.	Troch.	Fem.	Tibia.	Tarsi.	Claws.
Maple....................	96	60	175	135	105	
Honey-locust.............	108	30	180	135	78	27
Walnut....................	90	60	147	129	66	24
"	108	30	180	150	84	25
White elm...........		63	174	159	90	21

Pulvinaria pruni, n. sp. Plate XVI, fig. 3; plate XVII; plate XVIII, figs. 1, 2.

Scale of female. Before the formation of the ovisac the scale is not unlike that of the fully mature *Lecanium hesperidum,* of delicate texture, plane surface, oval, 1-1.7 mm. in width, 1.5-2.9 mm. long. After oviposition the scale becomes more dense, recurved, plicate, when boiled in KOH and spread out under cover glass measures about 3 mm. in width and 4 mm. in length. Marginal spines simple. The base of ovisac ranges from 5 to 7 mm. in length and from 3 to 5 mm. in width.

The larvæ settle on the twigs and both sides of the leaves, preferably the under side, in either case choosing positions alongside the veins. Longitudinal median carina prominent, undisturbed by shriveling of the body in the dried specimens.

The following measurements will show the characteristic structure of legs and antennæ of the adult female:

	ANTENNAL SEGMENTS.							
	1	2	3	4	5	6	7	8
Scale on leaves	57	75	84	63	39	30	36	51
Scale on twigs..........	54	66	84	48	36	27	27	45

	SEGMENTS OF LEG.						
	Coxa.	Tro.	Femur.	Tibia.	Tarsus.	Claw.	Breadth of femur.
Scale on leaves:							
Cephalic leg..........	120	135	240	255	108	30	111
Median leg	165	180	300	285	114	45	111
Posterior leg..........	180	180	300	255	115	45	105
Scale on twig:							
Cephalic leg..........	135	150	249	270	96	24	96
Median leg	159	174	285	240	105	36	108
Posterior leg..........	150	150	276	246	108	30	93

The above measurements were so characteristic, differing essentially (being in most cases uniformly larger) from either material or description before me, that I sent mounts, specimens *in situ* and notes to Dr. Howard, for comparison with departmental collections. In a letter he says: "I have asked Mr. Pergande to give it a careful examination, and he reports that it is apparently an undescribed species. We have received it before, and it bears the biologic number in our collection '6222.' We have received it from Charleston and Florence, S. C., both in 1894."

In our own collections it bears the lot number 399, which refers to like number in accession book. It is here recorded as being received from Wichita, Kan., July 23, 1895. The accompanying letter stated that these insects have been infesting the trees for three or four years.

On the twigs of this same plum tree *Aspidiotus ancylus* and *Aspidiotus forbesi* were present. This is the second instance of the association in goodly numbers of these two species, the first being lot D, on crab-apple.[4]

A number of the scales of *P. pruni* showed the effects of parasites. A specimen was sent to Doctor Howard, who finds it to be *Coccophagus lecanii* Fitch.

Parlatoria pergandei Comstock. Plate XIX, figs. 1, 2.

This species is differentiated by Comstock from its nearest ally, *Parlatoria proteus* Curt., principally by the shape of scale of female:

---

4. This lot was discussed in this journal, vol. VIII, No. 1, p. 3. Mention is there made of the presence of another species besides *A. forbesi* upon the crab tree, but the determination of the second species as *A. ancylus* was not given at that time.

circular in *pergandei*, oblong in *proteus;* by the fourth lobe : present in *pergandei*, absent in *proteus*.

It is upon these characters that specimens in hand are determined as *pergandei*. Comstock's comparisons were probably made without specimens of *proteus* at hand, since he mentions in a foot-note Signoret's figures and description.[5] Later, however, Comstock speaks of receiving specimens of *proteus* from Signoret,[6] confirming his conception of this species.

The exact status of *pergandei*, however, does not seem to be fully settled. Professor Cockerell in his first check-list cites *pergandei* as a distinct species,[7] and in his first supplement it is located as a variety of *proteus*.[8] I have received it from Mr. Craw on orange from the type locality, Florida, labeled in agreement with Cockerell's supplement. Mr. C. L. Marlatt, who is now studying the genus, says "*pergandei* Comst. (merges into *proteus* Curt.)"[9] Doctors Berlese and Leonardi place Comstock's variety of *pergandei*, *cameliæ*, as *proteus*, var. *cameliæ*,[10] and other instances likewise might be cited.

With *pergandei* on orange branches from Florida (Craw), the *pergandei* under consideration on orange leaves and branches from a Lawrence greenhouse, *proteus* on *Pinus insignis* from Perth, Australia (Ckll.), and *proteus* on leaf of an orchid, Watagode. Ceylon (Green), before me, the following notes are made : Regarding the form of scale of female, I find "circular" scales among *proteus* and "elongate" scales among *pergandei*. No steadfast distinction either in shape or color of the female scale can therefore be noticed. Concerning the pygidium, the *proteus* on *Pinus* and the orchid show, in accordance with Comstock, the presence of plates in the location where the pointed fourth lobe is found in *pergandei*. *Proteus* further shows marginal undulations apparently independent of the lobes, the crests situated beneath (ventral aspect) the lobes and extending out about one-third the length of the lobes. These undulations are shown in the figure and are characteristic of the *proteus* on *Pinus*. They are not noticeable on the orchid insects.

Proteus, further, is not recorded, as far as I can ascertain, existing upon orange. I have received specimens in exchange on orange labeled *proteus*, but discriminations made upon the above basis showed the insects to be undoubtedly *pergandei*. The undulations along the posterior margin of *proteus* may be fairly constant; their presence in this one lot on *Pinus* does not warrant a statement of their perma-

---

5. Rep. Comm'r Agr. 1880, p. 327.
6. 2d Rep. Dept. Ent. Cornell Univ. 1883, p. 114.
7. Bull. Ill. St. Lab., vol. IV, p. 335, 1896.
8. Bull. Ill. St. Lab., vol. V, p. 397, 1899.
9. Marlatt, MSS.
10. Chermothoca Italia, Fascicola I, No. 2.

nence. The question then arises, Are the distinctions surrounding the fourth-lobe position of specific moment? In some genera they would not be. In *Parlatoria* all species are closely related, and hence distinctive structural characteristics, though slight, are of greater weight than in genera where distinctions are more marked. The presence of the fourth lobe, rudimentary, with papillar termination, then, should differentiate the species *P. pergandei* from the species *P. proteus* with its fourth rudimentary lobe showing plates extending beyond its caudal margin.

NOTES AND CORRIGIENDA.

Part I.

Aspidiotus obscurus Comstock as recorded was represented by only a few individuals upon one tree. I have since (October, 1899) found two black oak trees quite generally infested on both trunk and limbs. These trees are in a forest near Holton, Jackson county, Kansas.

On page 4, and wherever it occurs thereafter, the specific term *ancyclus* should be *ancylus*.

Part II.

Lecanium canadense Ckll. Occurred sparsely when found at time of collection, April, 1898. Last year the same conditions existed. This season it appears to be abundant upon elms in Lawrence and vicinity; in some cases assembled in clusters. I have found a species, alike in form, size, and color upon hickory, a suggestion for further observations upon the status of *caryæ* and *canadense*.

Lecanium cockerelli Hunter. This species has been found on plum in Nebraska (Bruner MSS.) I have found it on hickory this season (May) near Lawrence. The scale on hickory was well covered with a pruinose coating.

Lecanium macluræ being preoccupied, the term *Lecanium aurantiacum* is now offered in substitution.

Measurements of length and breadth of bodies of scales contain thereafter the abbreviation   , which should be *mm.*

COCCIDÆ OF KANSAS, IV.

Additional Species, Food-plants and Bibliography of Kansas Coccidæ, with Appendix on other Species Reported from Kansas.

BY S. J. HUNTER. With plate XX

A.—Additional Species.

Kermes pubescens Bogue. Plate XX, fig. 1.

On white oak, Lawrence, Douglas county.

Kermes nivalis King and Ckll. Plate XX, fig. 2.

On white oak, Lawrence, Douglas county.

Orthezia graminis Tinsley. Plate XX, figs. 3, 4.

On goldenrod (*Solidago* sp.), Blue Rapids, Marshall county. Mrs. S. G. Cady, collector.

B.—Food-plants of Kansas Coccidæ.

In order to understand the significance or importance of the food-plants of Coccidæ, or scale-insects, some knowledge of the life and habits of the insect is necessary. Scale-insects are plant parasites and locate themselves upon the bark or outer covering of the plants. They have long, slender beaks, which they are able to insert into the tissues of the plants and draw therefrom the plant juices. Some scale-insects choose but a single host-plant, and others seem to be able to subsist upon a very great variety of plants. This adaptability to various food-plants has much to do with the numbers of the several species in existence. It is evident that if a species of insect has to depend exclusively upon a single plant variety, the chances of life for this insect would decrease with a decrease in numbers of the host; while, on the other hand, scale-insects which have the power to adapt themselves to a number of plants have greater chances of life and better opportunities for numerical increase. In animal parasitism the parasite tends to increase as the host increases. The increase of the parasite, however, is generally in a greater ratio than the increase of the host, so that the parasite frequently becomes so numerous as to destroy or greatly curtail the increase of the host, and then the parasite must succumb likewise, or adapt itself to new conditions. Such relations between host and parasite exist to a certain extent between the scale-insects and their respective hosts. A study of the food-plants, therefore, of the scale-insects, becomes a matter of considerable importance in determining the continuation of a species and the possibilities of its numerical increase. The insects herein discussed have been found on certain food-plants in Kansas. They have likewise been found by other authorities on other food-plants in other parts of the globe. A record of each of these discoveries is given in the following pages:

Aspidiotus forbesi Johns.

Honey-locust, *Gleditschia tricanthos* (Johns.), Ckll., Proc. Nat. M., XIX, p. 738.
Peach, *Prunus*, Ckll., ibid., p. 740.
Apricot, *Prunus armeniaca*, Johns., Ent. News, p. 151 (1896).
Garden currant, *Ribes rubrum*, ibid.
Ash, Osborn, Proc. Iowa Acad. Sci., p. 229 (1897).
Crab-apple, Hunter, K. U. Quart., VIII., No. 1, p. 4 (Jan. 1899).
Pear, Johns., Ill. Sta. Lab. Nat. Hist., IV, p. 381.
Plum, ibid.
Apple, ibid.
Quince, ibid.
Currant, ibid.
Wild and cultivated cherry, *Amygdalus persica*, Leonardi, Riv. di Pat. Veg.,
      p. 43 (1897).
Acer fraxinus, ibid.
Staphylea trifoliata, ibid.

Aspidiotus ancylus Putnam.

Linden, Comstock, 2d Corn. Univ. Rept., p. 140.
Box-elder, *Negundo* sp., Ckll., Proc. Nat. M., XIX, p. 735.
Apricot, *Prunus armeniacum*, Ckll., Proc. Nat. M., XIX, p. 741.
Plum, *Prunus domestica*, in Santa Fé, N. M.. Ckll., ibid.
Black currant, *Ribes* sp., Ckll., Am. Nat., p. 731 (1895).
Oaks, Comstock, 2d Corn. Univ. Rept., p. 140.
Beech, ibid., p. 139.
Water locust, ibid., p. 140.
Ilex verticillata, Felt., Bull. N. Y. Mus., VI, No. 31, p. 617 (1900).
Hemlock, ibid.
Mountain ash, ibid.
Willow, Felt., Bull. N. Y. Mus., V, No. 23, p. 261 (1898).
Apple, ibid.
Elm, ibid.
Pear, Gillette and Baker, Colo. Agr. Exp. Sta., Bull. 31, Tech. Ser., No. 1, p. 128.
Black maple, Newell, Cont. Iowa St. Col. Agr., No. 3, p. 8.
Birch, ibid.
Snowball, ibid.
Gleditschia tricanthos, Ann. Mag. Nat. Hist., p. 323 (1898).
Quercus wrightii, Ckll., Can. Ent., vol. 28, p. 226 (1896).
Cottonwood, Gillette, Colo. Agr. Col. Ex. Sta., Bull. No. 38, 1898, p. 36.
Spirea arnicus, King, Can. Ent., p. 226, vol. —.*

---

* The separates of the Canadian Entomologist have no date marks. It has been, therefore, impossible to locate accurately all references as to recent numbers of this magazine, since the department numbers are at the bindery.

Honey-locust, King, Can. Ent., p. 226, vol. —.
Quince, ibid.
Maple, Putnam, Proc. Dav. Acad. Nat. Sci., vol. II, p. 346.
Peach, Comstock, U. S. Dept. Rept. Com. Agr., 1880, p. 59.
Osage orange, ibid.
Hackberry, ibid.
Bladder-nut, ibid.
Ash, ibid.
Chestnut, in U. S. Dept. Agr. Coll. Howard.
Cratægus, ibid.
Elagnus reflexa, ibid.
Lonicera, ibid.
Syringa, ibid.
Prunus pissardi, ibid.

Aspidiotus uræ Comst.

Grape, Comst., 2d Rept. Dept. Ent. Cor. Exp. Sta., p. 71, 1883.
Hickory, ibid.

Aspidiotus osborni Newell.

Quercus alba, Hunter, K. U. Quart., vol. VIII, No. 1, p. 6.
Ironwood, *Ostrya virginica*, Newell, Cont. Iowa St. Col. Agr., No. 3, p. 7.

Aspidiotus ulmi Johns.

White elm, *Ulmus americana*, Johns., Ill. St. Lab. Nat., vol. IV, art. 13, p. 388.
Slippery or Red elm, *Ulmus fulva*, Hunter, K. U. Quart., VIII, No. 1, p. 6.
Catalpa, Hunter, K. U. Quart., VIII, No. 1, p. 6.

Aspidiotus fernaldi Ckll., subsp. *albiventer*.

Maple, *Acer* sp., Hunter, K. U. Quart., VIII, No. 1, p. 7.

Aspidiotus obscurus Comst.

Willow oak, Comst., 2d Corn. Univ. Rept., p. 140.
Black oak, *Quercus* sp., Hunter, K. U. Quart., VIII, No. 1, p. 7.
Chestnut, Hunter, found in Miami county, Kansas, June, 1901.

Aspidiotus juglans-regiæ Comst.

Peach, *Prunus* or *Amygdalus persica*, Ckll., Proc. Nat. M., XIX, p. 740.
English walnut, Comst., 2d Corn. Univ. Rept., p. 61.
Prune, *Prunus* sp. (Ehrhorn), Ckll., Can. Ent., 1895, p. 260.
Crab-apple, Hunter, K. U. Quart., vol. VIII, No. 1, p. 8.
Pear, Comst., 2d Corn. Univ. Rept., p. 62.
Cherry, ibid.
Locust, ibid.
Ash, Fernald, Pa. Dept. Agr., Bull. No. 43, p. 20.
Currant, Osborn, Proc. Iowa Acad. Sci., vol. V, p. 230, 1897.

Aspidiotus perniciosus.

Cherry, Howard, Bull. 12, U. S. Dept. Agr., Div. Ent., p. 13.
English huckleberry, ibid.
Black walnut, ibid.
Japan walnut, ibid.
English willow, ibid.

Golden willow, Bull. 12, U. S. Dept. Agr., Div. Ent., p. 13.
Rocky Mountain dwarf cherry, ibid.
Flowering quince, ibid.
Japanese quince, ibid.
Strawberry, ibid.
Black currant, ibid.
Lombardy poplar, ibid.
Carolina poplar, ibid.
Golden-leaved poplar, ibid.
Silver maple, ibid.
Cut-leaved birch, ibid.
Mountain ash, ibid.
Milkweed, ibid.
Catalpa speciosa, ibid.
Actinidia, ibid.
Citrus trifoliata, ibid.
Red dogwood, ibid.
Snowball, *Viburnum*, ibid.
Juneberry, ibid.
Loquat, ibid.
Laurel, ibid.
Akebia, ibid.
White currant, Lochhead, Ont. Dept. Agr., p. 31, Mar. 1900.
White ash, ibid.
Ornamental birch, ibid.
Maple leaf, ibid.
Rhubarb, ibid.
Hemp, ibid.
Lamb's-quarters, ibid.
Garden knotweed, ibid.
Mustard, ibid.
Beggar-ticks, ibid.
Goose-foot, ibid.
Ragweed, ibid.
Sunflower, ibid.
Weeping willow, Smith, Rept. N. J. Agr. Col. Exp. Sta., p. 547, 1896.
Laurel-leaved willow, ibid.
Kilmarnock willow, ibid.
Linden, ibid.
English walnut, ibid.
Flowering currant, ibid.
Euonymus, ibid.
Gooseberry, ibid.
Persimmon, Ebenaceæ, ibid.
Acacia, Leguminoseæ, ibid.
Elm, ibid.
Osage orange, ibid.
Pecan, ibid.
Hickory, ibid.
Alder, ibid.
Chestnut, ibid.
Oak, ibid.

Sumac, Smith, Rept. N. J. Agr. Col. Exp. Sta., p. 547, 1896.
Grape, ibid.
Catalpa bignoniodes, (Howard), Ckll., U. S. Dept. Agr., Div. Ent., Tech. Ser., No. 6, p. 17.
Crab-apple, Ckll., U. S. Dept. Agr., Div. Ent., Tech. Ser., No. 6.
Bartlett pear, ibid.
Dwarf Duchess pear, ibid.
Pyrus japonica, ibid.
Satsuma plum, ibid.
Prunus pissardi, ibid.
Prunus maritimi, ibid.
Citrus albapanetatus, ibid.
Cottonwood, ibid.
European linden, ibid.
Apple, *Pyrus malus*, Comst., 2d Corn. Univ. Rept., p. 140.
Apricot, *Prunus armeniaca*, ibid.
Crab grass, Johnson, Bull. No. 57, Md. Agr. Exp. Sta., p. 61, 1898.
Peach, Ckll., Proc. Nat. M., XIX, p. 740.
Rose, Ckll., Amer. Nat., p. 726, 1895.
Almond, Fernald, Bull. 36, Mass. Agr. Col., p. 19.
Spirea, ibid.
Raspberry, ibid.
Hawthorn, ibid.
Cotoneaster, ibid.

Aspidiotus greenii Ckll.

Palm, *Howea belmoreana*, Hunter, K. U. Quart., VIII, No. 1, p. 11.*
Banana, Townsend, An. & Mag. Nat. Hist., ser. 7, III, p. 169.*
Cycas (Green), Ckll., Bull. Div. Ent., Tech. Ser., No. 6, p. 27.*
House palm, Ckll., Ent. Mo. Mag., XXXIV, pp. 184, 185, Aug. 1898.*
Palm, *Seaforthia elegans*, Ckll., Entom., XXXII, p. 93, Apr. 1899.*
Guava, Ckll. and Parrott, Industrialist, p. 277, May, 1899.*
Vine leaves, ibid.*

Aspidiotus hederæ Ball., var. *nerii* Bouche.

Madrone, *Arbutus menziesii*, Coquillett, Bull. 26, Div. Ent. U. S. Dept. Agr., p. 20.
Century plant, *Agave americana*, ibid.
Lilac, *Syringa vulgaris*, ibid.
Nightshade, *Solanum douglasii*, ibid.
China tree, ibid.
English ivy, ibid.
Oak, *Quercus agrifolia*, ibid.
Arborvitæ, *Thuja occidentalis*, on cones of, ibid.
Acacia, Comstock, 2d Rept. Corn. Univ. Exp. Sta., 1883, p. 13.
Cherry, ibid.
Currant, ibid.
Grass, ibid.
Clover, ibid.
Orange tree, ibid.
Lemon, ibid.
Maple, ibid.

---

* Those succeeded by a star were kindly furnished by Mr. Kotinsky, through Dr. Howard.

Melia, Comstock, 2d Rept. Corn. Univ. Exp. Sta., 1883, p. 13.
Oleander, ibid.
Plum, ibid.
Yucca, ibid.
Coprosma lucida, Maskell, Sca. Ins. N. Z., p. 45.
Corynocarpus lœvigatœ, ibid.
Orchids, Gardners' Chronicle, May 6, 1893, p. 548.
New Zealand flax, *Phormium tenax*, Comst., 2d Corn. Rept., p. 140.
Holly, Comst., ibid.
Boxwood, Comst., 2d Corn. Rept., p. 139.
Oleander (Ckll.), Gillette, Colo. Agr. Col. Ex. Sta., Bull. No. 31, Tech. Ser., No. 1,
    p. 128.
Orange (Ckll.), Gillette, ibid.
Lemon (Ckll.), Gillette, ibid.
Fracaena (Ckll.), Gillette, ibid.
Olive, *Olea europea*, Ckll., Jour. Trin. Nat. Fld. Club, vol. II, No. 12, p. 307.
Magnolia grandiflora, ibid.
Rose, Townsend, Tech. Ser., No. 4, Div. Ent., p. 11.
Palo dulce, ibid.
Lace fern, Johnson, Div. Ent., n. ser., No. 6, pp. 75–78 (1896).
Sweet lime (fruit), Cockerell, Ann. Mag. Nat. Hist., p. 167, Feb. 1899.
Pinus (leaves), ibid.
Ivy, *Hedera helix*, Saccardo, Rivieta Pat. Veg., IV, p. 50, 1895 (1895-'96).
Nerium oleander, ibid, p. 49.
"Trueno," Cockerell, Biol. Cent. Amer., II, part 2, p. 20, Dec. 1899.
Palms, King, Can. Ent., XXXI, p. 225, Aug. 1899.
Cycas, ibid.
Heather, *Erica*, ibid.
Cycas revoluti, Osborn, Proc. Iowa Acad. Sci., V, p. 230, 1897 (1898).
Arica, Cockerell, Jour. Inst. Jamaica, I, p. 255, No. 55, Apr. 1893.
Tea, Green, Indian Mus. Notes, IV, No. 1, p. 4, 1896 (reprint, p. 3).
Loranthus, ibid.
Dalbergia, ibid.
Aristea major, Indian Ent. Mo. Mag., XXXIII, p. 69, Mar. 1897.
Aucuba, Douglae, Ent. Mo. Mag., XXIII, pp. 151, 152, Dec. 1896.
Azalea, ibid.

Aspidiotus œsculi Johns., subsp. *solus*, subsp. nov. Hunter.

Juglans nigra, Hunter, K. U. Quart., VIII, No. 1, p. 12.

Mytilaspis pomorum Bouche.

Apple, Comst., 2d Corn. Univ. Rept., p. 139.
Basswood, Ckll., Proc. Nat. M., XI, p. 730.
Bladder-nut, *Staphylœa* (Comst.), ibid., p. 736.
Broom, *Cytisus scoparius*, Ent. Mo. Mag., p. 138, 1893.
Cultivated locuet, *Robinia pseudacacia*, Ckll., Proc. Nat. M., XIX, p. 738.
Hawthorn, *Cratœgus oxycantha*, ibid., p. 744.
Wild gooseberry, *Ribes cynosbati*, Country Gentleman, p. 27, Jan. 16, 1895.
Cornus californicus (Harvey), Ckll., Proc. Nat. M., XIX, p. 735.
Pear, Maskell, Scale Insects of N. Z., p. 52 (1887).
Plum, ibid.
Peach, ibid.

Apricot, Maskell, Scale Insects of N. Z., p. 52 (1887).
Lilac, ibid.
Thorn, ibid.
Sycamore, ibid.
Cottoneaster, ibid.
Foliis variegatus (Harvey), Ckll., Proc. Nat. M., XIX, p. 752.
Cornus alba (Harvey), ibid.
Fraxinus americanus, Country Gentleman, p. 27, Jan. 10, 1895.
Planera (Comst.), Ckll., Proc. Nat. M., XIX, 766.
Yucca, Comst., 2d Corn. Univ. Rept., p. 140.
Poplar, Bull. N. Y. Mus., 13th Rept. St. Ent. p. 374, 1897.
Magnolia, umbrella, ibid.
Privet, Bull. N. Y. Mus., V, No. 23, p. 261, Dec. 1898.
Quince, Hunter, K. U. Bull. Dept. Ent., p. 25, 1898.
Hop-tree, Hunter, ibid.
Buckthorn, ibid.
Raspberry, ibid.
Grape-vines, Berlese and Leonardi, Rivista Pat. Veg., III, p. 347, 1895 (1894-'95).*
Citrous plants, ibid.*
Linden, Comstock, Rept. U. S. D. A., 1880, pp. 325, 326, pl. xix, fig. 2.*
Amorpha (an exotic), ibid.
Water locust, ibid.
Raspberry, ibid.
Ribes alpcrum, ibid.
Lonicera pulverulenla, ibid.
Planera kaku, ibid.
Pennsylvania maple, Hopkins, Can. Ent., XXVII, p. 248, 1896.*
Cornus californicus, Ckll., Can. Ent., XXVII, pp. 259, 260, 1895.*
Viburnum, King, Can. Ent., XXXI, p. 228, Aug. 1899.*
Spiræa aruncus, ibid.
Cornus alternifolia, ibid.
Ailanthus glandulosus, Maskell, Ent. Mo. Mag., XXXIII, p. —, Nov. 1897.*
Stillingia scbifera, ibid.
Common broom, *Sarothamnus scoparius*, Newstead, Ent. Mo. Mag., XXIX,
    p. 138, 1893.*
Cytisus nubigenus, ibid.*
Cocoa-palm leaves, Morgan, Ent. Mo. Mag., XXV, p. 350, Aug. 1889.*
Currant, Hunter, Bull. Dept. Ent., p. 25, Jan. 1898.
Horse-chestnut, ibid.
Maple, ibid.
Water locust, ibid.
Honeysuckle, ibid.
Elm, ibid.
Hackberry, ibid.
Cottonwood, ibid.
Willow, ibid.
Wild red cherry, Lochhead, Ont. Dept. Agr., p. 41, Mar. 1900.
Grape, ibid.
Spirea, ibid.
Juneberry, ibid.
Birch, ibid.
Rose, ibid.

Bittersweet, Lochhead, Ont. Dept. Agr., p. 41, Mar. 1900.
Walnut, Smith, N. J. Agr. Exp. Sta., p. 140.
Butternut, ibid.
Sycamore (Maskell), Ckll., Proc. Nat. Mus., vol. XIX, p. 736.

Diaspis snowii Hunter.

Salix nigra, Hunter, K. U. Quart., VIII, No. 1, p. 15.

Lecanium (Maclura) aurantiacum Hunter.

Osage orange, *Maclura aurantiaca*, Hunter, K. U. Quart., VIII, No. 2, p. 68.

Lecanium canadense Ckll.

Ulmus racemosa, Ckll., Can. Ent., p. 254, 1895.
Ulmus americana (Glenn), Hunter, K. U. Quart., VIII, No. 2, p. 69.

Lecanium kansasense Hunter.

Cersis canadensis, Hunter, K. U. Quart., VIII, No. 2, p. 69.
Juglans nigra, Hunter, K. U. Quart., VIII, No. 2, p. 70.
Ulmus fulva, Hunter, ibid.

Lecanium cockerelli, nov. sp., Hunter.

Ulmus americana, Hunter, K. U. Quart., VIII, No. 2, p. 70.
Juglans-nigra, Hunter, K. U. Quart., VIII, No. 2, p. 71.
Oak, *Quercus* sp., King, Can. Ent., vol. —, p. 252.
Sweet fern, *Comptonia asplenifolia*, ibid.
Plum, Hunter, K. U. Quart., vol. IX, Ser. A, No. 2, p. 107, Apr. 1900.
Hickory, ibid.
Populus sp., King, Psyche, IX, p. 117, Oct. 1900.
Wild crab-apple on University campus, Hunter.

Lecanium armeniacum Craw.

Plum, Hunter, K. U. Quart., VIII, No. 2, p. 71.
Grape-vine, Country Gentleman, June 16, 1898.
English gooseberry, Felt., Bull. N. Y. Mus., VI, No. 31, p. 617.
Apricot, Webster, Can. Ent., XXX, No. IV, 1898.
Prune, ibid.
Cherry, ibid.
Pear, ibid.
Peach, Cockerell, Jour. N. Y. Ent. Soc., VII, p. 257, Dec. 1899.*

Lecanium hesperidum Linnæus.

Pittosporum (Coq.), Ckll., Proc. Nat. M., XIX, p. 727.
Tea plant, *Camellia*, ibid., p. 728.
Abutilon, Gillette and Baker, ibid., p. 729.
Citrus tree, Comst., 2d Corn. Univ. Rept., p. 139.
Holly, Ckll., Proc. Nat. M., XIX, p. 733.
Grape-vine, Coq., Bull. 26, Div. Ent. U. S. Dept. Agr., p. 26.
Rhus integrifolia (Coq.), Ckll., Proc. Nat. M., XIX, p. 736.
Schinus molle, ibid., p. 737.
Cultivated locust, *Robinia pseudacacia*, ibid., p. 738.
Apricot, Coq., Bull. 26, Div. Ent. U. S. Dept. Agr., p. 26.
Rose, ibid.

Loquat, Coq., Bull. 26, Div. Ent. U. S. Dept. Agr., p. 26.
Ash, ibid.
Box, *Buxus sempervirens*, Maskell, Sca. Ins. N. Z., p. 111.
Willow, Coq., Bull. 26, Div. Ent. U. S. Dept. Agr., p. 26.
Lombardy poplar, ibid.
Orchid, *Dendrobium*, Ckll., Trans. Amer. Ent. Soc. 1893, p. 49.
Hippeastrum equestre, Ckll., Proc. Nat. M., XIX, p. 777.
Palm, Ckll., Ins. Life, VI, p. 103.
Calla lily, Coq., Bull. 26, Div. Ent. U. S. Dept. Agr., p. 26.
English ivy, *Hedera helix*, Hunter, K. U. Quart., VIII, No. 2, p. 75.
Oleander, Comst., 2d Corn. Univ. Rept., p. 140.
Rubber tree, *Ficus macrophylla*, Coq., Bull. 26, U. S. Dept. Agr., Div. Ent.,
      p. 26.
Euonymus, ibid.
Maple, *Acer dasycarpum*, ibid.
Rhamus crocea, ibid.
Heteromeles arbutifolia, ibid.
Fig, ibid.
House fern, N. Y. Mus., 13th Rept., p. 374, 1897.
English laurel, Felt., Bull. N. Y. Mus., V, No. 23, p. 260.
Aralia, Gillette and Baker, Colo. Agr. Exp. Sta., Bull. 31, Tech. Ser., No. 1, p. 127.
Ficus clastica, ibid.
Rhynchospermum jasimoides, ibid.
Veronica hendersonii, Gillette and Baker, Colo. Agr. Exp. Sta., Bull. 31, Tech.
      Ser., No. 1, p. 127.
"Fitolaca," Townsend, Bull. 4, Tech. Ser., Div. Ent. U. S. D. A., pp. 11 and 13,
      1896.*
Guava, ibid.*
Maitenus boaria, Cockerell, Act. Soc. Sci. Chili, V, p. XXIV, 1895.*
Areca catechu, Ckll., Insect Life, V, p. 159, 1893.*
Pepper tree, Riley and Howard, Insect Life, IV, p. 294, 1892.*
Bertolonia marchanti, Green, Ent. Mo. Mag., XXXIII, p. 71, Mar. 1897.*
Lucuma multiflora, Green, ibid.*
Dalbergia lanceolaria, ibid.*
Hedera amurensis, ibid.*
Carica papaia, Maskell, Ent. Mo. Mag., XXXIII, p. 243, Nov. 1897.*
"Ohia," ———, ibid.*

Lecanium coffeæ Walker.

Sword fern, *Pteris*, Hunter, K. U. Quart., VIII, No. 2, p. 75.
Tea plant, *Camellia*, Cotes, Ind. Mus. Notes, 1895.
Coffee, *Coffea* sp., Ckll., Bull. Bot. Dept. Jamaica, p. 71, 1894.
Orange, King, Can. Ent., p. 140, vol. —.
Diospyros, ibid.
Oleander, ibid.
Chrysophyllum, ibid.
Sago palm, ibid.
Croton variegatum, ibid.
Cycas revoluta, Ckll. and Parrott, Industrialist, May, 1899, p. 276.*
Psidium, Hempel, Riv. Mus. Paulista, IV, p. 426, 1900.*
Gardenia, Atkinson, Jour. Asiat. Soc. Bengal, LV, pt. II, No. 3, pp. 282-284,
      1886.*
Gardenia florida, Maskell, Ent. Mo. Mag., XXXIII, p. 243, Nov. 1897.*

Lecanium olcæ Bernard.

Camellia, *Camellia japonica*, Ckll., Proc. Nat. M., XIX, p. 529.
"Brachæton," meaning perhaps Brachychiton, ibid., p. 730.
Lignum-vitæ, *Guaiacum officinale*, ibid., p. 731.
Holly (Coq.), Ckll., ibid., p. 733.
Euonymus sp. (Coq.), Ckll., ibid., p. 734.
Grape-vine, *Vitis inconstans*, Ins. Life, 1893, p. 160.
Sycamore (Coq.), Ckll., Proc. Nat. M., XIX, p. 736.
Schinus molle, in Mexico (Coq.), ibid., p. 737.
Apricot, Comst., 2d Corn. Univ. Rept., p. 140.
Rose, ibid.
Guava (Coq.), Ckll., Proc. Nat. M., XIX, p. 748.
Pomegranate, *Punica granatum*, Ckll., ibid., p. 750.
Cape jessamine, *Gardenia jasinoides* (Comst.), Ckll., ibid., p. 752.
Artemisia californica (Coq.), Ckll., ibid., p. 754.
Ash, Coq., Bull. 26, Div. Ent. U. S. Dept. Agr., p. 28.
Lingustrum lucidum, Coq., Bull. 26, Div. Ent. U. S. Dept. Agr., p. 28.
Red pepper, ibid.
Phoradendron flavescens (Johns.), Ckll., Proc. Nat. M., XIX, p. 764.
In Calif. *Enfagus* sp., Coq., Bull. 26, Div. Ent. U. S. Dept. Agr., p. 28.
Cycas revoluta, Coq., ibid., p. 29.
Hippeastrum equestre, in Mexico, Ckll., Ins. Life, V, p. 245.
Palm, Comst., 2d Corn. Univ. Rept., p. 140.
Oleander, *Nerium oleander*, Coq., Bull. 26, Div. Ent. U. S. Dept. Agr.
Photinia or *Heteromeles arbutifolia*, ibid, p. 28.
Citrus, Comst., 2d Corn. Univ. Rept., p. 140.
Olive, ibid.
Live oak, Bull. Bot. Dept. Jamaica, p. 12, 1894.
Peas, ibid.
Plum, Ckll., Bull. Bot. Dept. Jamaica, p. 12, 1894.
Bittersweet, ibid.
Apple, ibid.
Eucalyptus sp., ibid.
Almond, ibid., p. 72.
Acer dasycarpum, ibid.
Artemisia californica, ibid.
Abutilon, ibid.
Rhus integrifolia (Coq.), ibid.
Baccharis virminalis, ibid.
Ficus macrophylla, ibid.
Habrothamnus elegans, ibid.
Irish juniper, ibid.
Myosporum, ibid.
Melalenca purpurea, ibid.
English laurel, ibid.
Beech, ibid.
Rhamnus crocea, ibid.
Grevillca robusta, ibid.
Ligustrum japonicum, ibid.
Indian cedar, ibid.
Cedar of Lebanon, ibid.
Castor-bean, ibid.

Sonchus oleraeens, Ckll., Bull. Bot. Dept. Jamaica, p. 12, 1894.
Abutilon, Coq., Bull. 26, Div. Ent. U. S. Dept. Agr., p. 29.
Naphitum litsehii, Osborn, Proc. Iowa Acad. Sci., p. 226, 1897.
Lombardy poplar, Coq., Bull. 26, Div. Ent. U. S. Dept. Agr., p. 23.
Pepper tree, ibid.
Solanum jasimoides, Gillette and Baker, Bull. 31, Colo. Exp. Sta., p. 127, May,
　　1895.*
Platycerium aleieornr, ibid.*
Eriodendron, Ckll., Appdx. Bull. Misc. Information (Trinidad), II, pp. 3 and 4,
　　No. 23, Apr. 1896.*
Fern, Ckll., Amer. Nat., XXIX, p. 727, Aug. 1895.*
Cassinia ceptophylla, Maskell, XVII, p. 28, 1884 (1885).*
Calabassa tree, Busek., Bull. 22, n. s., Div. Ent. U. S. D. A., p. 92, 1900.*
Honey-loeust, ibid.*
Guazuma ulmifolia, ibid.*
Terminalia eatappa, ibid.*
Meyenia alba, Ckll., Insect Life, V, p. 160, 1893.*
Mango ——, ibid.*
Yucea, Ckll., Can. Ent., XXVII, p. 257, Sept. 1895.*
Deciduous magnolia, Craw., Bull. 4, Tech. Ser., Div. Ent. U. S. D. A., p. 40., ft.-
　　note, Apr. 1896.*
Pelargonium, Townsend, Bull. 4, Tech. Ser., Div. Ent. U. S. D. A., pp. 11 and
　　13, 1896.*
"Marguerita," ibid.*
Catalpa, ibid.*
Croton eluteria, Green, Ent. Mo. Mag., XXXIII, p. 72, fig. 2, Mar. 1897.*
Brexia, ibid.*
Avicennia nitida, ibid.*
Eleodendron orientale, ibid.*
Carissa spinarum, ibid.*
Catesbœa spinosa, ibid.*
Aralia elegantissima, ibid.*
Macrozamia frazeri, ibid.*

Lecaniodiaspis (?) parrotti Hunter.

Æsculus glabra, Hunter, K. U. Quart., VIII, No. 2, p. 76.

Lecaniodiaspis celtidis Ckll., subsp. *pruinosus* Hunter.

Ulmus americana Hunter, K. U. Quart., VIII, No. 2, p. 77.

Chionaspis ortholobis Comst.

Willow, *Salix* sp., Comst., 2d Corn. Univ. Rept., p. 140.
Cottonwood, *Populus* sp., Hunter, K. U. Quart., IX, No. 2, p. 101.
Populus grandidentata King, Psyche, IX, p. 117, Oct. 1900.

Chionaspis salicis-nigrae Walsh.

Ash, Comst., 2d Corn. Univ. Rept., p. 139.
Poplar (Osborn), Cooley, Spec. Bull. Mass. Agr. Coll., p. 21, Aug. 1899.
Willow, Cooley, Spec. Bull. Mass. Agr. Col., p. 21, Aug. 1899.
Cottonwood, ibid.
Salix alba, ibid.
Salix alba, var. *camellia*, ibid.

Salix nigra, Cooley, Spec. Bull. Mass. Agr. Col., p. 21, Aug. 1899.
Cornus pubescens, ibid.
Cornus asperifolia, ibid.
Balm of Gilead, ibid.
Russian poplar, ibid.
Liriodendron tulipifera, ibid.
Cornus stolonifera, ibid.
Cornus sericea, ibid.
Ceanothus, ibid.
Amelanchier canadensis, ibid.

Chionaspis americana Johns.

Ulmus americana (Johns.), Ckll., Proc. Nat. M., XIX, p. 765.

Chionaspis platani Cooley.

Sycamore (Parrott), Cooley, Spec. Bull. Hatch Exp. Sta., Aug. 1899, p. 36.

Chionaspis pinifoliæ Fitch.

Pine, Comst., 2d Corn. Rept., p. 140, 1880.
Firs and spruces, Gillette and Baker, Hemiptera of Colorado, p. 129.
Pinus strobus, Cooley, Spec. Bull. Hatch Exp. Sta., Aug. 1899, p. 33.
Pinus resinosa, ibid.
Pinus excelsa, ibid.
Pinus mitis, ibid.
Pinus cembra, ibid.
Pinus pyrenaica, ibid.
Pinus laricis, ibid.
Pinus sylvestris, ibid.
Pinus austriaca, ibid.
Pinus punilio, ibid.
Pseudotsuga taxifolia, ibid.
Abies excelsa, ibid.
Abies nigra, ibid.
Abies alba, ibid.

Pulvinaria innumerabilis Rathv.

Tilia sp., Ckll., Proc. Nat. M., XIX, p. 730.
Euonymus (Riley), Ckll., Proc. Nat. M., XIX, p. 734.
Vitis inconstans, also on wild grape, Ckll., Proc. Nat. M., XIX, p. 735.
Sycamore (Riley), Ckll., ibid., p. 736.
Rosa sp., Riley, Rept. Dept. Agr., 1884.
Garden currants, *Ribes rubrum* and *nigrum*, Riley, Rept. Dept. Agr., 1884.
Fraxinus nigra (*Sambricifolia*), Mundt, Can. Ent., 1884, p. 240.
Poplars, *Populus balsamifera*, ibid.
Mulberry, *Morus rubra*, ibid.
Oak, *Quercus* sp., Riley, Rept. Dept. Agr., 1884.
White elm, *Ulmus americana* Hunter, Q. U. Quart., IX, No. 12, p. 104.
Black walnut, *Juglans nigra*, ibid.
Maple, ibid.
Silver maple, Putnam, Dav. Acad. Nat. Sci., p. 294.
Beech, ibid.
Sugar maple, ibid.
Box-elder, ibid., p. 338.

Locust, Putnam, Dav. Acad. Nat. Sci., p. 338.
Sumac, ibid.
Soft maple, Fitch, Trans. N. Y. St. Agr. Soc. 1859, XIX, pp. 775, 776.
Æsculus flava, King, Psyche, p. 154, Jan. 1901.
Virginia creeper, ibid.
Tilia curopœa, Mann, Psyche, IV, p. 221, 1884.
Robinia pseudacacia, ibid.
Vitis labrusca, ibid.
Rhus glabra, ibid.
Vitis riparia, ibid.
Fagus, ibid.
Salix, ibid.
Maclura, ibid.
Quercus, ibid.
Ulmus, ibid.
Platanus, ibid.
Ribes, ibid.
Euonymus, ibid.
Celtio, ibid.
Morus rubra, Howard, Bull. 22, n. s., Div. Ent. U. S. Dept. Agr., p. 8, 1900.
Aralia japonica, ibid.
Viburnum dentatum, King, Psyche, IX, p. 117, Oct. 1900.
Strawberry, Forbes, 13th Ann. Rept. Insects Ill., p. 98, 1883 (1884).

Pulvinaria pruni Hunter.

Plum, Hunter, K. U. Quart., IX, No. 2, p. 105.

Parlatoria pergandii Comst.

Orange, *Citrus* sp., Comst., 2d Corn. Univ. Rept., p. 140.
Tangerine, Felt., Bull. N. Y. St. M., VI, No. 31, p. 618.
Date palm, King, Can. Ent., p. 228.
Variegated croton, Ckll., Jour. Inst. Jamaica, I, p. 55, Feb. 1892.*

Chionaspis furfura Fitch.

Apple, Cooley, Hatch Exp. Sta. Mass. Agr. Col., Sp. Bull., p. 28, 1899.
Cherry, ibid.
Currant, ibid.
Japan quince, ibid.
Crab-apple, ibid.
Mountain ash, ibid.
European mountain ash, ibid.
Black walnut, ibid.
Black-cap raspberry (Osborn), ibid.
Rhamnus catharticus (King), ibid.
Clethra alnifolia (King), ibid.
Pyrus arbutifolia (Kirkland), ibid.
Pyrus nigra (Kirkland), ibid.
Pyrus hetrophylla (Kirkland), ibid.
Pyrus salicifolia pendula (Kirkland), ibid.
Pyrus floribunda (Kirkland), ibid.
Pyrus spectabilis (Kirkland), ibid.
Pyrus pinnatofilia (Kirkland), ibid.
Ribes sanguinem (Morgan), ibid.

Choke cherry, *Prunus virginiana* (Howard), King, Psycho, vol. VIII, 1899, p. 335.
Black cherry, *Prunus serratina*, ibid.
Wild red cherry, *Prunus pennsylvanica*, ibid.
Pear, *Pyrus communis*, ibid.
Peach, *Persica vulgaris*, ibid.
Cherry currant, *Ribes* sp., ibid.
Red flowering currant, *Ribes sanguinem*, ibid.
European mountain ash, *Larbus anicuparia*, ibid.
Flowering quince, King, ibid.
Shadbush, *Amelanchier canadensis*.
Black alder, *Clethra alnifolia*, ibid.
Amelanchier canadensis, King, Psyche, IX, pp. 117, 118, Oct. 1900.
Populus grandidentata, ibid.

Mytalaspis citricola Packard.

Citrus trees, Ckll., Proc. Nat. Mus., vol. XIX, p. 732.
Banksia integrifolia (Maskell), Ckll., ibid., 763.
Kroton (Maskell), Ckll., ibid., p. 765.
Toddalia aculeata, Green, Coccidæ of Ceylon, part I, pp. 78-80, pl. XX, 1896.
Myrraya, Cockerell, Insect Life, p. 160, 1893*.
Coccutus indicus, Green, Indian Mus. Notes, IV, No. 1, p. 4, 1896 (reprint, p. 3).*
Tangerine, Riley and Howard, Insect Life, VI, p. 51, 1893. *
Taxus cuspidata, Mask., Ent. Mo. Mag., XXXIII, p. 241, Nov. 1897. *
Quercus, ibid.*

Kermes pubescens Bogue.

Scrub oak, Bogue, Can. Ent., vol. 30, p. 172.
White oak, King, *Quercus* sp., Can. Ent., vol. 30, p. 139.

Kermes niralis King and Ckll.

Quercus alba, King, Can. Ent., vol. 30, p. 139.

Orthesia graminis Tinsley.

Culms and blades of grass, Tinsley, Can. Ent. vol. 30, p. 13, 1898.
Goldenrod, *Solidago* sp., Hunter, K. U. Quart., vol. X, A, p. 107.*

C.—Bibliography.

Aspidiotus forbesi.

Aspidiotus forbesi Johnson, Ent. News, 7, No. 5, pp. 150–152, (1896).
do   Johnson, Ill. Lab. Nat. Hist., vol. 4, p. 380, (1896).
do   Johnson, Dept. Agr., Div. Ent., Bul. 6, n. ser., pp. 75–78, (1896).
do   Forbes, 20th Ann. Rept. Insects Ill., pp. 15, 16, 1895, (1896).
do   Cockerell, Dept. Agr., Div. Ent., Tech. Ser., No. 6, p. 21, 1897.
do   Johnson, Dept. Agr., Div. Ent., n. s., Bul. No. 9, pp. 83–85, (1897).
do   Leonardi, Rivistadi Patologia Vegetale, p. 49, (1897-1900).
do   Parrott, Trans. Ks. St. Hort. Soc., vol. 23, pp. 108, 109, (1898).
do   Hunter, Bull. Dept. Ent., p. 24, (1898).
do   Popenoe, Trans. Ks. St. Hort. Soc., vol. 23, pp. 40–46, (1898).

Aspidiotus forbesi Leonardi, Annali di Agr., p. 110, (1898).
do   Hopkins, Dept. Agr., Div. Ent., Bull. 17, n. s., pp. 144-149, (1898).
do   Johnson, Can. Ent., p. 82, (Apr. 1898).
do   Cooley, Dept. Agr., Ent. Bull. 17, n. s., pp. 61-65, (1898).
do   Brick, Sta. fur Pflanzenschutz zu Hamburg, 1, p. 34, (1898-'99).
do   Reh, Sta. fur Pflanzenschutz zu Hamburg, 1, p. 19, (1898-'99).
do   Cockerell, Jour. N. Y. Ent. Soc., vol. VII, p. 258, (Dec. 1899).*
do   Newell, Cont. from Ent. Dept. Iowa Agr. Col., No. 3, p. 14, (1899).
do   Reh, Mitth. Naturhist. Mus. Hamburg, vol. XVI, pp. 125-141, Mar. 1899.
do   Hunter, K. U. Quart., vol. 8, No. 1, p. 3, (Jan. 1899).
do   King, Can. Ent., vol. XXXI, p. 226, (1899).
do   Lochhead, Ont. Dept. Agr., p. 35, (Mar. 1900).
do   Newell, Iowa Sta.; Bull. 43, pp. 145-176, (1900).
do   Frank & Kruger, Schildlausbuch, pp. 74, 75, fig. 37, (1900).*
do   Reh, Zeitschrift fur Entom., vol. V, pp. 161, 162, June, 1900.*

Aspidiotus ancyclus.

Aspidiotus ancyclus Fitch,† Ann. Rept. N. Y. St. Agr. Soc., vol. XVI, p. 426, (1856).
Diaspis ancyclus Putnam, Tr. Iowa St. Hort. Soc., vol. XII, p. 321, (1877).
Aspidiotus ancyclus Putnam, Proc. Dav. Acad., vol. II, p. 346 (1878).
do   Comstock, Rept. Ent. in Rept. Com. Agr., p. 292, (1880).
do   Comstock, 2d Rept. Corn. Exp. Sta., p. 58, (1888).
do   Cockerell, Ins. Life, vol. VII, p. 210, (1894).
do   Cockerell, Can. Ent., vol. XXVI, p. 191, (1894).*
do   Cockerell, Bull. 2, n. s., U. S. Dept. Agr., p. 94, (1895).
do   Cockerell, Can. Ent., vol. XXVII, pp. 16, 17, (1895).*
do   Lugger, Minn. Sta., Bull. 43, pp. 99 *et seq.*, (1895).
do   Gillette and Baker, Colo. Agr. Col. Exp. Sta., Bull. No. 31, Tech. Ser., No. 1, p. 128, (1895).
do   Cockerell, Garden and Forest, vol. VIII, p. 513, (1895).*
do   Lowe and Sirrine, N. Y. Sta. Rept., pp. 525-535, (1896).
do   Webster, Ind. Hort. Rept., p. 9, (1896).
do   Johnson, Dept. Agr., Div. Ent., Bull. 6, n. s., pp. 75-78, (1896).
do   Forbes, 20th Ann. Rept. Insects Ill., pp. 15, 16, (1896).*
do   Hopkins, Bull. 6, n. s., Div. Ent. U. S. Dept. Agr., p. 72, (1896).
do   Cockerell, Can. Ent., vol. XXVIII, p. 226, (1896).*
do   Barrows and Pettit, Mich. Sta., Bull. 160, pp. 334 *et seq.*, (1897).
do   Cockerell, Dept. Agr., Div. Ent., Tech. Ser., No. 6, p. 20, (1897).
do   Leonardi, Rivist. Pat. Veg., p. 42, (1897).
do   Leonardi, ibid, pp. 216-218, (1897).
do   King, Can. Ent., vol. ——, p. 226, (——).

---

† As *A. circularis*, n. sp., but according to C. L. Marlatt MS. is *A. ancyclus.*

Aspidiotus ancyclus Johnson, Can. Ent., vol. XXX, p. 82, (1898).
do   Cooley, Dept. Agr., Div. Ent., Bull. 17, n. s., pp. 61-65, (1898).
do   Gillette, Colo. Agr. Col. Exp. Sta.,ǀBull. No.38, p. 36, (1898).
do   Parrott, Trans. Kan. St. Hort. Soc., vol. XXIII, pp. 108-109, (1889).
do   Leonardi, Annali di Agr., p. 108, (1898).
do   Osborn, Proc. Iowa Acad. Sci., vol. V, p. 229, (1898).*
do   Cockerell, Ann. Mag. Nat. Hist., p. 323, (1898).*
do   Goethe, Berl. Kgl. Lehraustalt Obst. Wein., p. 19, (1898).*
do   May, Station fur Pflanzenschutz zu Hamburg, vol. I, p. 5, (1898-'99).
do   Reh, ibid, p. 19, (1898-'99).
do   Brick, ibid, p. 31, (1898-'99).
do   Fernald, Com. Pa. Dept. Agr., Bull. 43, p. 20, (1899).
do   Newell, Cont. from Ent. and Zool. Dept. Iowa St. Agr. Col. No. 3, p. 7, (1899).
do   Hunter, K. U. Quart., vol. VIII, No. 1, p. 4, (1899).
do   Troop, Ind. Sta., Bull. 78, pp. 45-52, (1899).
do   Newell, Iowa Sta., Bull. 43, pp. 160-162, (1899).
do   Marlatt, Can. Ent., vol. XXXI, pp. 208-211, (1899).*
do   Reh, Illustr. Zeitschr. f. Ent., vol. IV, p. 246, (1899).*
       Mitth. Nat. Hist. Mus. Hamb., vol. XVI, pp. 125-141, (1899).*
do   May, ibid., pp. 151-153, (1899).*
do   Reh, Zeitschrift, fur Ent., vol. V, pp. 161, 162, (1900).*
do   Reprint Jahrs. Hamb. Anst., vol. XVII, pp. 4 *et seq.*, (1900).
do   Frank and Kruger, Schildlausbuch, pp. 72, 73, (1900).*
do   Lochhead, Ont. Dept. Agr., p. 37, (1900).
do   Felt., Bull. N. Y. Mus., vol. VI, No. 31, p. 579, (1900).
do   Marlatt, Ent. News, vol. XI, pp. 590-592, (1900).*
do   Scott, Bull. 26, n. s., Div. Ent. Dept. Agr., p. 50, (1900).
do   Reh, Bull. 22, n. s., Div. Ent. Dept. Agr., pp. 79-83, (1900).

Aspidiotus uvæ.

Aspidiotus uvæ Comstock, Rept. Ent. in Rept. Com. of Agr., pp. 309, 310, (1880).
do   Murtfeldt, Ent. Rept. Dept. Agr., pp. 135, 136, (1888).*
do   Cockerell, Jour. Inst. Jamaica, p. 142, (1892).*
do   Cockerell, Insect Life, vol. IV, p. 333, (1892).*
do   Cockerell, Jour. Inst. Jamaica, p. 255, No. 52, (1893); Jour. Trinidad Field Nat. Club, 1, p. 312, No. 88, (1894).*
do   Howard, Insect Life, vol. VII, p. 53, (1894).*
do   Riley, Md. Sta. Rept., pp. 190, 191, (1894).
do   Cockerell, Garden and Forest, vol. VIII, p. 513, (1895).*
do   Webster, Ind. Hort. Rept., p. 16, (1896).
do   Johnson, Dept. Agr., Div. Ent., Bull. VI, n. s., pp. 75-78, (1896).
do   Cockerell, Dept. Agr., Div. Ent., Tech. Ser., No. 6, p. 23, (1897).
do   Leonardi, Rivista di Patalogia Vegetale, vol. VI, p. 44, (1897).
do   Parrott, Trans. Kan. Hort. Soc., vol. XXIII, pp. 106-109, (1898).
do   Chambliss, Tenn. Sta. Bull., vol. X, No. 4, pp. 141-151, (1898).
do   Hunter, K. U. Quart., vol. VIII, No. 1, p. 4, (1899).

Aspidiotus uvœ Newell, Cont. from Ent. Dept. Iowa Agr. Coll. No. 3, p. 12, (1899).
do   Scott, Bull. XXVI, n. s., Div. Ent. U. S. Dept. Agr., p. 50,(1900).*

Aspidiotus osborni.

Aspidiotus osborni Newell, Cont. from Ent. Dept. Iowa St. Agr. Coll., No. 3, p. 5, (1895).
do   Osborn, Proc. Iowa Acad. Sci., V, p. 229, (1897).*
do   Hunter, K. U. Quart., vol. VIII, No. 1, p. 5, (1899).
do   Newell, Iowa Sta., Bull. 43, p. 165, (1899).
do   Scott, Bull. 26, n. s., Div. Ent. U. S. Dept. Agr., p. 50, (1900).*

Aspidiotus ulmi.

Aspidiotus ulmi Johnson, Ill. St. Lab. Nat., vol. IV, No. 13, p. 388, (1896).
             U. S. Dept. Agr., Div. Ent., Bull. 6, n. e., p. 75, (1896.
             Ent. News, vol. VII, p. 152, (1896).*
do   Cockerell, Bull. VI, Tech. Ser., Div. Ent. U. S. Dept. Agr., p. 22, (1897).
             Can. Ent., vol. XXIX, p. 266, (1897).
             Bull. Ill. St. Lab. Nat. Hist., vol. V, art. VIII, p. 396, No. 1026, (1899).
do   Newell, Cont. from Ent. Dept. Iowa Ag. Col. No. 3, p. 28, (1899).
do   Hunter, K. U. Quart., vol. VIII, No. 1, p. 6, (1899).

Aspidiotus fernaldi.

Aspidiotus fernaldi albiventer, Hunter, K. U. Quart., vol. 8, No. 1, p. 6, (1899).
do   *cockerelli,* Parrott, Can. Ent., vol. 31, No. 1, p. 10, (1899).
do   *albiventer,* Newell, Cont. from Ent. Dept. Iowa St. Agr. Col. No. 3, p. 19, (1899).

Aspidiotus obscurus.

Aspidiotus obscurus Comstock, Rept. Ent. in Rept. Com. of Agr., p. 303, (1880).
do   Forbes, 20th Ann. Rep. Ins. Ill., pp. 15, 16, (1895).*
do   Cockerell, Dept. Agr., Div. Ent., Tech. Ser., No. 6, pp. 9 and 21, (1897).
do   Leonardi, Rivista Pat. Veg., vol. VII, pp. 205-207, (1898).*
do   Hunter, K. U. Quart., vol. VIII, No. 1, p. 7, (1899).
do   Scott, Bull. XXVI, n. s., Div. Ent. U. S. Dept. Agr., p. 50, (1900).*

Aspidiotus juglans-regiæ.

Aspidiotus juglans-regiæ Comstock, Rept. Ent. in Rept. Com. Agr., p. 300, (1880).
do   Colvee, Bull. Soc. Ent. Fr., ser. 6, vol. I, p. CLXV,* (1881). (Describes it as a juglandis.)
do   Comstock, 2d Rept. Corn. U. Exp. Sta., p. 61, (1883).
do   Howard, U. S. Dept. Yearbook, pp. 251-252, 254, 264, 265, (1894).
do   Riley and Howard, Ins. Life, vol. VI, p. 328, (1894).*
             Can. Ent., vol. 26, No. 2, p. 36, (1894).
             Can. Ent., vol. 26, p. 131, (1894).

Aspidiotus juglans-regiæ Cockerell, Proc. U. S. Nat. Mus., vol. XVII, pp. G15-
G25, (1895).
Garden and Forest, VIII, p. 513, 25, (1895).*
Bull. Ill. St. Lab. Nat. Hist., vol. IV, art. XI,
p. 333, No. 551, p. 334, (1896).*
do   Lintner, Bull. 6, n. s., Div. Ent. U. S. Dept. Agr., p.
60, (1896).*
do   Johnson, Dept. Agr., Div. Ent., Bull. VI, pp. 75-78,
(1896).
do   Cockerell, Div. Ent. U. S. Dept. Agr., Tech. Ser.,
Bull. VI, pp. 18-21, (1897).
do   Starnes, Ga. Sta., Bull. 36, p. 31, (1897).
do   Hillman, Nev. Sta., Bull. 36.
do   Osborn, Proc. Iowa Acad. Sci., V, p. 230, (1897).*
do   Leonardi, Riv. di Pat. Veg., vol. VI, p. 67, (1897).
Riv. di Pat. Veg., vol. VII, 40, (1898).*
do   Cockerell, Bull. St. Lab. Nat. Hist., vol. V, art. VII,
p. 396, No. 589, (1899).*
Bull. 32, Ariz. Exp. Sta., p. 282, (1899).*
do   Newell, Cont. from Ent. Dept. Iowa St. Agr. Col.
No. 8, p. 19, (1899).
do   Fernald, Com. Pa. Dept. Agr., Bull. No. 43, p. 19,
(1899).*
do   Hunter, K. U. Quart., vol. VIII, No. 1, p. 8, (1899).
do   Lochhead, Ont. Dept. Agr., p. 38, (1900).
do   Scott, Bull. 26, n. s., Div. Ent. U. S. Dept. Agr., p.
50, (1900).*
do   Morgan, La. Sta., Bull. 28, 2d ser., pp. 282-1005,
(18 - -).*

Aspidiotus perniciosus.

Aspidiotus perniciosus Comstock, Rept. Comr. Agr. 1880, pp. 304, 305, (1880).
do   Cooke, Treatise on Ins. Inj. to Frt. and Frt. Trees, (1881).
do   Riley, Rept. U. S. Dept. Agr., Div. Ent., p. 65, (1881).*
do   Chapin, Rept. Comr. Agr., pp 207, 208, (1882).
1st Rept. St. Bd. Hort. Cal., pp. 65-68, (1882).*
Reprint Pacif. Rur. Press, pp. 2-5, (1882).*
do   Cooke, Rept. Comr. Agr., pp. 65, 208, (1882).
Ins. Inj. to Orchards, pp. 60-63, (1883).
do   Chapin, Rept. Cal. St. Bd. Hort., p. 91, (1883).
Bull. No. 2, St. Bd. Hort., (1884).
do   Klee, Rept. Cal. St. Bd. Hort., pp. 373-375, 1885-86,
(1887); ibid., p. 404, (1881).
do   Klee, Hatch, Buck, *et al.*, discussion, ibid., (1887).
do   Boggs, ibid., p. 12, (1887).
do   Rose, Brittan, and Mileo, ibid., pp. 40, 41, (1887).
do   Klee, Rept. St. Bd. Hort., 1887-'88, p. 245, (1888).
do   Klee, Treatise on Ins. Inj. to Frt. and Frt. Trees of
Cal., pp. 10, 11, (1888).
do   Cooper, Pacific Rur. Press, pp. 146, 147, (1889).
do   Coquillett, Weekly Blade, (1890).
do   Washburn, Bull. 5, Ore. Agr. Exp. Sta., p. 23, (1890).

Aspidiotus perniciosus Lelong, Rept. Cal. Sta. Bd. Hort., 1889, p. 170, (1890).
do   Riley and Howard, Insect Life, vol. III, p. G9, (1890).
do   Kercheval and Gray, Insect Life, vol. II, p. 312, (1889).
do   Freeman, Insect Life, vol. III, p. 68, (1890).
do   Riley, Rept. U. S. Dept. Agr., Div. Ent., p. 262, (1890).*
do   Gregorson, Insect Life, vol. III, p. 169, (1890).
do   Collins, Rept. Cal. St. Bd. Hort., 1890, (1890).
do   Coquillett, Bull. 23, Div. Ent. Dept. Agr., pp. 19-36, (1891).
do   Allen, Bull. No. 2, 1st Bien. Rept. Ore. St. Bd. Hort., 1891, pp. 43-46, (1891).
do   Riley and Howard, Insect Life, vol. III, p. 426, (1891 .
do   Insect Life, vol. III, p. 487, (1891).
do   Insect Life, vol. IV, p. 83, (1891).
do   Lelong, Rept. St. Bd. Hort. Cal. for 1891, (1892).
do   Craw, Rept. Cal. St. Bd. Hort. for 1891, (1892).
do   Coquillett, Rept. on Sca. Ins. of Cal., (1891), Bull. 26, Div. Ent., pp. 21-25, (1892).
do   Craw, Rept. Cal. St. Bd. Hort. 1891, p. 285, (1892).
do   Riley, Rept. Eut. for 1891, p. 244, (1892).
do   Townsend and Tyler, Bull. 7, N. M. Agr. Exp. Sta., pp. 6, 7, (1892).
do   Olliff, Agr. Gazette, N. S. Wales, pp. 298, 299, (1892)
do   Riley and Howard, Insect Life, vol. V, p. 53, (1892).
do   Riley, Insect Life, vol. V, p. 127, (1892).
do   Koebele, Albert, Craw, Rept. Importation Parasites and Pred. Ins., p. 15, (1892).
do   Coquillett, Insect Life, vol. V, p. 251, (1893).
do   Riley and Howard, Insect Life, vol. V, p. 280, (1893).
do   Cockerell, Ann. Mag. Nat. Hist., p. 406, (1893).*
do   Allen, Bull. No. 5, 2d Rept. Bien. Rept. Ore. St. Bd. Hort., pp. 67-69, (1893).
     Bull. No. 6, 2d Bien. Rept. Ore. St. Bd. Hort., pp. 83-86, (1893).
do   Cockerell, Bull. No. 10, N. M. Agr. Exp. Sta., pp. 14-16, (1893).
do   Koebele, Insect Life, vol. VI, p. 26, (1893).
do   Special Bull. U. S. Dept. Agr., p. 39, (1893.)
do   Ehrhorn, Rept. Cal. St. Bd. Hort., 1893-'94, p. 375, (1894).
do   Cockerell, Can. Ent., vol. XXVI, p. 32, (1894).
do   Smith, Insect Life, vol. VII, pp. 185-186, (1894).*
     Insect Life, vol. V, p. 312, (1894).*
do   Coquillett, Insect Life, vol. VI, p. 253, (1894).
do   Schwarz, Insect Life, vol. VI, p. 247, (1894).
do   Riley and Howard, Insect Life, vol. VI, p. 207, (1894).
do   Howard, Circ. No. 3, 2d ser., Div. Ent., 10 pp., (1894).
do   Lawson, Chronicle (Halifax, N. S.), (1894).
do   Lelong, Pacific Rural Press, (1894).
do   Riley and Howard, Insect Life, vol. VI, p. 286, (1894).
do   Coquillett, Insect Life, vol. VI, p. 324, (1894).
do   Smith, Ent. News, vol. VII, pp. 182-184, (1894).
     Garden and Forest, vol. VII, p. 344, (1894).

Aspidiotus perniciosus Riley, Rept. Ent. for 1893, (1894).
  do   Riley and Howard, Insect Life, vol. VI, p. 360, (1894).
  do   Riley, Proc. Amer. Adv. Sci., vol. XLIII, p. 229, (1894).
  do   Howard, Insect Life, vol. VII, p. 153, (1894).
  do   Smith, Insect Life, vol. VII, p. 163, (1894).
  do   Aldrich, Insect Life, vol. VII, p. 201, (1894).
  do   Cockerell, Insect Life, vol. VII, p. 207, (1894).
  do   Sirrine, Garden and Forest, vol. VII, p. 449, (1894).
  do   Smith, Bull. 106, N. J. Exp. Sta., p. 24, (1894).
  do   Lintner, Albany Eve. Jour., (1894).
  do   Howard, Can. Ent., vol. XXVI, p. 353, (1894).
  do   Lintner, Rural New Yorker, vol. LIII, p. 791, (1894).
  do   Thorne and Webster, Ohio Exp. Sta., Emerg. Poster
                  Supp. Bull., p. 56, (1894).
  do   Fletcher, Evidence before Com. on Agr., p. 19, (1894);
                  25th An. Rept. Ent. Soc. Ontario, pp. 73-76, (1894).
  do   Smith, Rept. Ent. Dept. N. J. Agr. Col. Exp. Sta. for
                  1894, (1895).
  do   Webster, Bull. 56, Ohio Exp. Sta., for 1894 (1895).
  do   Weed, Bull. No. 23, N. H. Agr. Exp. Sta., (1894).
  do   Smith, Ent. News, p. 312, (1894).
  do   Webster, Proc. Columbus, Ohio, Hort. Soc., pp. 168-169,
                  (1894).
  do   Riley, Rept. Va. Sta. Bd. Hort. Soc., pp. 172-178, (1894).
  do   Rolfs, Proc. 7th Ann. Meet. Fla. Hort. Soc., pp. 94-99,
                  (1894).
  do   Howard, Year-book U. S. Dept. Agr. for 1894, pp. 249-
                  276, (1895).
  do   Fletcher, Rept. Exptl. Farms Canada for 1894, pp. 183-
                  226, (1895).
  do   Webster, Ohio Farmer, p. 157, (1895).
  do   Boaz, So. Planter, p. 119, (1895).
  do   Sirrine, Bull. No. 87, N. Y. Sta. Agr. Exp. Sta., (1895).
  do   Webster, Entomology, (1895).
  do   Howard, Proc. Ent. Soc. Wash., vol. III, pp. 210-226
                  (1895).
  do   Sturgis and Britton, Bull. No. 121, Conn. Agr. Exp. Sta.,
                  pp. 6-14, (1895).
  do   Sturgis, 19th Ann. Rept. Conn. Agr. Exp. Sta., for 1895,
                  (1896).
  do   Washburn, Bull. No. 38, Ore. Agr. Exp. Sta., pp. 7, 8,
                  (1895 .
  do   Webster, Ohio Farmer, p. 315, (1895).
  do   McCarthy, Bull. No. 120, N. C. Agr. Exp. Sta., (1895).
  do   Hillman, Bull. No. 29, Nev. Agr. Exp. Sta., p. 8, (1895).
  do   Piper, Bull. No. 17, Wash. Agr. Exp. Sta., (1895).
  do   Schiedt, Rept. Pa. Sta. Bd. Agr., pp. 579-584, (1895).
  do   Gillette and Baker, Bull. 31, Colo. Exp. Sta., p. 128, (1895).
  do   Maskell, Agr. Gazette, N. S. Wales, vol. VI, pp. 868-
                  870, (1895).
  do   Cockerell, Can. Ent., vol. XXVII, pp. 16, 17, (1895); Bull.
                  2, n. s., Div. Ent. U. S. Dept. Agr., p. 94, (1895).

Aspidiotus perniciosus Giard, Act. Soc. Sci. Chili, vol. V, p. 147, (1895).
do   Smith, Ent. News, vol. VI, pp. 153-157, (1895).
do   Slingerland, Rural New Yorker, vol. LVI, p. 5, (1895).
do   Fox and Edge, Circ. Pa. St. Bd. Agr., (1895).
do   Howard, Insect Life, vol. VII, pp. 283, 333, 359, 360,
               (1895).
do   Collingwood, Rural New Yorker, vol. LIV, p. 167, (1895).
do   Collier, Bull. 87, n. s., N. Y. Agr. Exp. Sta., p. 10, (1895).
do   Riley, Bull. 32, Md. Agr. Exp. Sta., (1895).
do   Davis and Taft, Bull. 121, Mich. Agr. Exp. Sta., pp. 36-
               38, (1895).
do   Lintner, Bull. N. Y. St. Mus., vol. III, No. 13, pp. 273-
               303, (1895).
do   Hoffman, The Student, (1895).
do   "Rather Perplexed" (pseud.), So. W. Farm and Or-
               chard, (1895).
do   Smith, Ent. News, vol. VI, pp. 153-157, (1895).
do   Beckwith, Bull. 25, Del. Coll. Agr. Exp. Sta., p. 8, (1895).
do   Toumey, Bull. 14, Ariz. Agr. Exp. Sta., pp. 32-47, (1895).
do   Marlatt, Insect Life, vol. VII, pp. 365-374, (1895).
do   Cockerell, New Mexican, (1895).
do   Rolfs, Bull. 29, Fla. Agr. Exp. Sta., p. 19, (1895).
do   Fernald, Mass. Crop Rept., p. 23, (1895).
do   Maskell, Can. Ent., p. 14, (1896).
do   Sirrine, Ann. Rept. N. Y. Agr. Sta. for 1895, (1896).
do   Smith, Rept. Ent. Dept. N. J. Agr. Coll. Exp. Sta. for
               1895, (1896).
do   Garman, 8th Ann. Rept. Ky. Agr. Exp. Sta., for 1895,
               (1896).
do   Beckwith, Bull. No. 30, Del. Agr. Exp. Sta., p. 16, (1896).
do   Cooley, Bull. No. 36, Mass. Hatch Agr. Exp. Sta., pp.
               13-20, (1896).
do   Cordley, Bull. No. 44, Ore. Agr. Exp. Sta., p. 108, (1896).
do   Weed, So. Cultivator, (1896).
do   Alwood, Bull. 62, Va. Agr. Sta., pp. 31-44, (1896).
do   Cockerell, Bull. No. 19, N. M. Agr. Exp. Sta., pp. 108-
               112, (1896).
do   Hopkins, Nat. Stockman and Farmer, p. 6, (1896).
do   Cook, Rural Cal., pp. 158, 159, (1896).
do   Stedman, Mo. Month. Crop Rept., (1896).
do   Kinney, Bull. No. 37, R. I. Agr. Exp. Sta., p. 43, (1896).
do   Alwood, Bull. No. 66, Va. Agr. Exp. Sta., pp. 77-90,
               (1886).
do   Webster, Bull. No. 72, Ohio Agr. Exp. Sta., pp. 211-217,
               (1896).
do   Smith, Bull. No. 116, N. J. Agr. Exp. Sta., p. 15,
               (1896).
do   Johnson, Bull. No. 42, Md. Agr. Exp. Sta., pp. 154-156,
               (1896).
do   Smith, Bull. No. 6, n. s., Div. Ent. U. S. Dept. Agr.,
               pp. 46-48, (1896).

Aspidiotus perniciosus Lintner, Bull. No. 6, n. s., U. S. Dept. Agr., pp. 54-61, (1896).

do   Johnson, Bull. No. 6, n. s., Div. Ent. U. S. Dept. Agr., pp. 63-66, (1896).

do   Webster, Bull. No. 6, n. s., Div. Ent. U. S. Dept. Agr., pp. 66-70, (1896).

do   Hopkins, Bull. No. 6, n. s., Div. Ent. U. S. Dept. Agr., pp. 71-73, (1896).

do   Alwood, Bull. No. 6, n. s., Div. Ent. U. S. Dept. Agr., pp. 80-84, (1896).

do   Howard, Trans. Mass. Hort. Soc., p. 15, (1896).

do   Beckwith, Trans. Penin. Hort. Soc., pp. 85-90, (1896).

do   Webster, Ann. Rept. Ohio Sta. Hort. Soc., pp. 164-178, (1896).

do   Lintner, 11th Rept. on Inj. Ins. St. N. Y., pp. 200-233, (1896).

do   Slingerland, Proc. West. N. Y. Hort. Soc. for 1896, p. 13, (1896).

do   Smith, Bull. 6, n. s., Div. Ent. U. S. Dept. Agr., pp. 46-48, (1896).*

do   Green, Ent. Mo. Mag., vol. XXXII, p. 84, (1896).*

do   Hopkins, Bull. 6, n. s., Div. Ent. U. S. Dept. Agr, pp. 71, 72, (1896).*

do   Maskell, Ent. Mo. Mag., vol. XXXII, pp. 33-36, (1896).*

do   Forbes, 20th Ann. Rep. Ins. Ill., pp. 1-25, (1895-'96).*

do   Alwood, Bull. 6, n. s., Div. Ent. U. S. Dept. Agr., pp. 80-81, (1896).*

do   Johnson, Bull. 6, n. s., Div. Ent. U. S. Dept. Agr, pp. 63-65, (1896).*

do   Webster, Bull. 6, n. s., Div. Ent. U. S. Dept. Agr., pp. 69, 70, (1896).*

do   Maskell, Trans. N. Z. Inst., vol. XXVIII, p. 386, (1896).

do   Howard and Marlatt, Bull. No. 3, n. s., Div. Ent. U. S. Dept. Agr., pp. 1-30, (1896).

do   Rolfs, Rept. Fla. Agr. Exp. Sta. for 1896, (1897).

do   Smith, Rept. N. J. St. Bd. Agr., 24 pp., (1897).*

do   Berlese and Leonardi, Riv. Pat. Veg., vol. VI, pp. 330-352, (1897).*

do   Coquillett, Rept. Dept. Agr., Div. Ent., p. 121, (1897).*

do   Webster, Proc. 52d Ann. St. Agr. Conv., .Columbus, Ohio, (1897).

do   Sturgis, Conn. Agr. Sta. Rept. for 1896 (1897).

do   Baker, Bull. No. 77, Ala. Agr. Exp. Sta., pp. 27-31, (1897).

do   Alwood, Proc. Ga. St. Hort. Soc., pp. 38-42, (1897).

do   Forbes, Bull. No. 48, Agr. Exp. Sta., pp. 413-428, (1897).

do   Gillette, Bull. No. 38, Colo. Agr. Exp. Sta., pp. 33-39, (1897).

do   Alwood, Rept. 22d Meet. Amer. Ass. of Nurserymen, pp. 25-32, (1897).

do   Garman, Bull. No. 67, Ky. Agr. Exp. Sta., pp. 43-59, (1897).

Aspidiotus perniciosus Slingerland, Rural New Yorker, p. 356, (1897).
do   Rolfs, Garden and Forest, pp. 217, 218, (1897).
do   Hussey, Ohio Farmer, p. 487, (1897).
do   Webster, Can. Ent., p. 173, (1897); Bull. No. 81, Ohio
                 Agr. Exp. Sta., pp. 177-212, (1897).
do   Cockerell, Bull. No. 6, Tech. Ser. Div. Ent. U. S. Dept.
                 Agr., p. 31, (1897).
do   McCarthy, Bull. No. 138, N. C. Agr. Exp. Sta., pp. 45-
                 55, (1897).
do   Forbes, Circular, (1897).
do   Baker, Bull. No. 86, Ala. Agr. Exp. Sta., (1897).
do   Rolfs, Bull. 41, Fla. Agr. Exp. Sta., pp. 519-542, (1897).
do   Pantou, Rural Canadian, pp. 178, 179, (1897).
do   Smith, Circ. N. J. Agr. Exp. Sta., (1897).
do   Starnes, Bull. No. 36, Ga. Agr. Exp. Sta., p. 31, (1897).
do   Smith, Ent. News, pp. 221-223, (1897).
do   Alwood, 1st Ann. Rept. Insp. San Jose Sc., 15 pp., (1897).
do   Cordley, Ore. Agr. & Rural N. W., p. 70, (1897).
do   Barrows, Bull. No. 9, n. s., Div. Ent. U. S. Dept. Agr.,
                 pp. 27-29, (1897).
do   Webster, Bull. No. 9, n. s., Div. Ent. U. S. Dept. Agr.,
                 pp. 40-45, (1897).
do   Johnson, Bull. No. 9, n. s., Div. Ent., U. S. Dept. Agr.,
                 pp. 80-82, (1897).
do   Troop, Newspaper Bulletin, (1897).
do   Webster, Ent. News, pp. 248-250, (1897).
do   Massey, So. Planter, p. 549, (1897).
do   Hopkins, W. Va. Farm Reporter, pp. 84-86, (1897).
do   Osborn, Bull. No. 36, Iowa Agr. Exp. Sta., pp. 860-864,
                 (1897).
do   Berlese and Leonardi, Riv. Pat. Veg., vol. VII, pp. 252-
                 273, (1898).*
do   Leonardi, Riv. Pat. Veg., vol. VII, pp. 189-191, (1898).*
do   Osborn, Proc. Iowa Acad. Sci., vol. V, p. 321, (1898).*
do   Symposium, Smith, Lowe, Waugh, Bailey, Willard,
                 Lyon, Hale, and Bancroft, Rural New Yorker, pp.
                 17, 18, (1898).
do   Smith, E. N. Y. Horticulturist, (1898).
do   Johnson, Bull. No. 17, n. s., Div. Ent. U. S. Dept. Agr.,
                 pp. 39-43, (1898).*
do   Alwood, Bull. No. 72, Va. Agr. Exp. Sta., p. 11, (1898).
do   Cockerell, Entom., vol. XXXI, p. 239, (1898).*
                 Ent. News, vol. IX, pp. 95, 96, (1898).*
do   Smith, Bull. No. 125, N. J. Agr. Exp. Sta., pp. 9-14,
                 (1898).
do   Howard and Marlatt, Bull. No. 3, n. s., Div. Ent. U. S.
                 Dept. Agr., p. 80, (1898).
do   Gould, Bull. No. 144, Corn. Univ. Agr. Exp. Sta., pp.
                 9-11, (1898).
do   Newstead, Entom., vol. XXXI, p. 98, (1898).*
do   Discussion, Proc. Ent. Soc. London, p. 13, (1898).*

Aspidiotus perniciosus Smith, Rept. Ent. Dept. N. J. Agr. Col. Exp. Sta. for
1897, pp. 436–492, (1898).

do   Perkins, Bull. No. 60, Vt. Agr. Exp. Sta., pp. 12–14,
(1898).

do   Forbes, 20th Rept. St. Ent. Ill., pp. 1–25, (1898).

do   Emory, Trans. Penin. Hort. Soc., pp. 107-113, (1898).

do   Bancroft, Trans. Penin. Hort. Soc., pp. 113-128, (1898).

do   Webb, Trans. Penin. Hort. Soc., pp. 114, 115, (1898).

do   Smith, Bull. No. 17, n. s., Div. Ent. U. S. Dept. Agr.,
pp. 32–39, (1893).*

do   Fletcher, Rept. Ent. Soc. Ontario, vol. XXVIII, pp.
78, 79, (1898).*

do   Johnson, Can. Ent., vol. XXX, pp. 82, 83, (1898).*

do   Webster, Can. Ent. vol. XXX, pp. 169–172, (1898).*

do   Felt., Bull. 17, n. s., Div. Ent. U. S. Dept. Agr., pp. 22,
23, (1898).*

do   Hopkins, Bull. 17, n. s., Div. Ent. U. S. Dept. Agr.,
pp. 44, 45, (1898).*

do   Discussions, Bull. 17, n. s., Div. Ent. U. S. Dept. Agr.,
pp. 50–55, (1898).*

do   Matrdorff, Zeits. fur Pflanzenkranheiten, vol. VIII, p.
1, (1898).*

do   Sorauer, Zeits. fur Pflanzenkrankheiten, vol. VIII, p.
46, (1898) *

do   Sorauer, Zeits. fur Pflanzenkheiten, vol. VIII, p. 104,
(1893).*

do   Stedman, Bull. No. 41, Mo. Agr. Exp. Sta., p. 35,
(1898).

do   Hunter, Bull. Dept. Ent. Kans. Univ., p. 62 (1898).

do   Howard, Bull. No. 12, n. s., Div. Ent. U. S. Dept. Agr.,
pp. 1-31, (1898).

do   Johnson, Can. Ent., vol. XXXI, pp. 87, 88, (1899).*

do   Scott, Bull. 20, n. s., Div. Ent. U. S. Dept. Agr., pp.
83, 84, (1899).*

do   d'Utro, Bull. Inst. Agron, S. Paulo, vol. X, Nos. 11
and 12, (1899).*

do   Goethe, Bericht Kgl. Lehraustadt Obst. Wein.and Gar-
tenbau, Geisenheim a R., p. 16, (1899).*

do   Newstead, Reprint Jour. Royal Hort. Soc., p. 8, (1899).*

do   Webster, Can. Ent., vol. XXXI, p. 4, (1899).*

do   King, Can. Ent., XXXI, pp. 225, 226, (1899).*

do   Howard and Marlatt, Bull. 20, n.s., Div. Ent.U. S. Dept.
Agr., pp. 36-39, (1899).

do   Howard, Year-book Dept. Agr., pp. 143, 144, (1899).

do   Johnson, Bull. No. 20, n. s., Div. Ent. U. S. Dept. Agr.,
(1899).*

do   Reh, Mitth. Nat. Mus. Hamburg, vol. XVI, pp. 125–
141, (1899).*

do   Cockerell, Bull. No. 32, Ariz. Agr. Exp. Sta., p. 274,
(1899).*

do   Hunter, Kans. Univ. Quart., vol. VIII, No. 1, pp. 10, 11,
(1899).

Aspidiotus perniciosus Cockerell and Parrott, Industrialist, p. 276, (1899).*
do   Fuller, Tr. Ent. Soc. London, pt. IV, p. 465, (1899).*
do   Lampa, Ent. Tidsk., pp. 63-65, (1899).*
do   Frank and Kruger, Schildlausbuch, pp. 58-72, (1900).*
do   Smith, Bull. No. 146, N. J. Agr. Exp. Sta., 20 pages, (1900).*
do   Scott, Bull. No. 26, n. s., Div. Ent. U. S. Dept. Agr., p. 49, (1900).*
do   Woodworth, Bull. No. 26, n. s., Div. Ent. U. S. Dept. Agr., p. 92, (1900).*
do   Reh, Bull. 32, n. s., Div. Ent. U. S. Dept. Agr., pp. 79-83, (1900).*
             Zool. Anzeiger, No. 624, pp. 502-504, (1900).*
             Reprint Jahrb. Hamburg wiss. Anst., pp. 1-21, (1900).*
             Zeitschrift fur Entom., vol. V, pp. 161,162, (1900).*
do   May, Mitth. Naturhist. Mus. Hamburg, vol. XVI, pp. 151-153, (1900).*
do   Johnson, Bull. No. 26, n. s., Div. Ent. U. S. Dept. Agr., pp. 73-75, (1900).*
do   Cockerell, Science, n. s., vol. XII, No. 300, pp. 488, 489, (1900).*
do   Kellogg, Science, n. s., vol. XIII, pp. 383-385, (1901).
do   Webster, Science, n. s., vol. XIII, p. 511, (1901).

Aspidiotus greenii.

Aspidiotus greenii Cockerell, Dept. Agr., Div. Ent., Tech. Ser., No. 6, p. 27, (1897).
do   Cal. Fruit Grower, 20, No. 23, pp. 4, 5, (1897).
do   Hunter, K. U. Quart., vol. VIII, No. 1, p. 11, (1899).
do   Cockerell, An. and Mag. Nat. Hist., ser. 7, III, p. 169, (1899).
             Am. Nat., XXXI, p. 73, Aug. 1897.*
             Ent. Mo. Mag., XXXIV, pp. 184, 185, Aug. 1898.*
             Entom., XXXII, p. 93, Apr. 1899.*
do   Ckll. and Parrott, Industrialist, p. 277, May, 1899.*
do   Marlatt, Can. Ent., XXXI, pp. 208-211, Aug. 1899.*
do   Hempel, Rev. Mus. Paulista, IV, p. 502, 1900.*

Aspidiotus hederæ nerii Bouche.

Aspidiotus nerii Bouche, Schadl. Gart. Ins., p. 52, (1833).
Diaspis bouchei Targioni-Tozzetti, Stud. sul. Coccin, (1867).
Aspidiotus nerii Comstock, Rept. Ent. iu Rept. Com. Agr., p. 301, (1880).
do   Maskell, Trans. N. Zeal. Inst., XIV, p. 217, (1881).*
do   Comstock, 2d Rept. Corn. Univ. Exp. Sta., p. 63, (1883).*
do   Hubbard, U. S. Dept. Agr., p. 35, (1885).
do   Oscar Schmidt, Archiv. Naturgesch., LI, No. 1, pp. 169-200, (1885).*
             Thesis at Berlin University, (1885); rev. in Ent. Nach., p. 119, (1886).*
do   Lemoine, Ann. Soc. Ent. France, p. CXC, (1886).*
do   Douglas, Ent. Mo. Mag., XXIII, pp. 151, 152, (1886).*

Aspidiotus nerii Maskell, Scale Insects N. Zeal., p. 44, (1887).*
do   Morgan, Ent. Mo. Mag., XXV, p. 350, (1889).*
do   Cockerell, Gard. Chron., p. 518, No. 14, (1893).*
          Journ. Inst. Jamaica, p. 255, (1893).*
          Trans. Am. Ent. Soc., XX, p. 367, (1893).
          Ent. News, vol. V, pp. 79, 211, (1894).*
          Ent. Mo. Mag., XXX, pp. 57–59, (1894).*
          N. M. Ent., No. 2, (1894).
          Journ. Trin. Nat. Field Club, vol. 2, No. 12, p. 307,
            (1896).
do   Johnson, Div. Ent., n. s., No. VI, pp. 75–78, (1886).
do   Townsend, U. S. Dept. Agr., Div. Ent., Tech. Ser., No. 4, p. 11,
          (1896).
do   Green, Indian Mus. Notes, vol. IV, No. 1, p. 4, (1896).*
          Ent. Mo. Mag., XXXIII, p. 69, (1897).*
do   Gillette, Colo. Agr. Coll. Exp. Sta., Bull. No. 31, Tech. Ser.,
          No. 1, p. 129, (1898).
do   Hunter, K. U. Quart., vol. VIII, p. 11, (1899).
do   Newell, Iowa Sta., Bull. 43, p. 168, (1899).
do   King, Psyche, vol. VIII, No. 276, p. 250, (1899).
do   Cockerell, Am. Mag. Nat. Hist., p. 167, (1899).*
          Biol. Cent. Am., II, part II, p. 20, (1899).*

Aspidiotus æsculi solus (s. sp. n.)

Aspidiotus æsculi solus Hunter, K. U. Quart., vol. VIII, No. 1, p. 12, (1899).
do   Newell, Cont. from Ent. Dept. Iowa St. Agr. Coll. No.
          3, p. 13, (1899).

Mytilaspis pomorum.

Aspidiotus conchiformis of Anthers, but not A. *conchiformis* Gmelin, Syst.
          Nat., 2221, 37 (1788), which species infests elm.
do   *pomorum* Bouche, Ent. Zeitung Stett. (1851), 12, No. 1.
do   *pyrus-malus* Rob. Kennicot, Acad. Sci. Cleveland, (1854).
do   *juglandis* Fitch, Ann. Rept. N. Y. St. Agr. Soc., p. 163, (1856).*
Mytilaspis pomicorticis Riley, 5th Rept. St. Ent. Mo., p. 95, (1872).
Mytilaspis pomorum, Signoret, Ann. de la Soc. Ent. de France, p. 95, (1870).
do   Maskell, Tr. N. Zeal. Inst., vol. XI, pp. 192–194, (1879).*
Mytilaspis (pomicorticis) pomorum Comstock, Rept. Ent. in Rept. Com. Agr.,
          (1880).
        do   Signoret, Tr. on Ins. Inj. Fruits and Frt.
               Trees, by M. Cooke, Sacr., p. 37,
               (1881).
        do   Bethune, Ann. Rept. Ent. Soc. Ont., pp.
               77, 78, (1881).
        do   Comstock, 2d Rept. Dept. Ent., Corn.
               Univ. Exp. Sta., p. 118, (1883).
        do   Douglas, Ent. Mo. Mag., vol. XXII, p. 68,
               (1884).*
               Douglas, Ent. Mo. Mag., vol. XXII, p.
               219, (1886).*
               Douglas, Ent. Mo. Mag., vol. XXIII, p.
               27, (1886).*

Mytilaspis (*pomicorticis*) *pomorum* Maskell, Sca. Ins. in N. Z., p. 51, (1887).
   do   Weed, Bull. Ohio Agr. Exp. Sta., vol. III, No. 4, p. 127, (1890).
   do   Maskell, Ent. Mo. Mag., vol. XXVI, p. 277, (1890).*

Mytilaspis pomorum Fletcher, Can. Cent. Exp. Farm, Bull. No. 11, p. 36, 1891.
   do   Townsend, N. M. Sta., Bull. No. 3, p. 19, (June, 1891).
   do   Panton, Ont. Agr. Coll. Exp. Sta., Bull. No. 62, p. 7, (1891).
   do   Huet and Louise, Bull. Min. Agr., Paris, 11, No. 7, pp. 765-768, (1892).
   do   Smith, Rept. Ent. N. J. Agr. Coll. Exp. Sta., p. 494, (1894).
   do   Howard, Year-book U. S. Dept. Agr., pp. 252, 254, 259, (1894).*
   do   Howard, U. S. Dept. Year-book, pp. 249-276, (1894).
   do   Lintner, Bull. N. Y. St. Mus. 3, No. 13, pp. 276-305, (1895).
   do   Taft, Mich. St. Agr. Coll. Exp. Sta., Bull. No. 121, p. 25, (Apr. 1895).
   do   Lowe and Sirrine, N. Y. St. Rept., pp. 525-535, (1896).
   do   Saccardo, Rivista Pat. Veg., IV, p. 54, (1896).*
   do   Berlese, Le Cocciniglie Italiane, P. III, p. 297, (1896).
   do   Berlese, Rivista Pat. Veg., IV, Nos. 1-12, (1896).*
            Rivista Pat. Veg., V, Nos. 1-4, (1896).*
   do   Webster, Ind. Hort. Rept., p. 6, (1896).
   do   Johnson, Dept. Agr., Div. Ent., Bull. 6, n. s., pp. 75-78, (1896).
   do   Fletcher, Can. Exp. Farms Repts., pp. 250-264, (1896).
   do   Harvey, Mo. Sta. Rept., pp. 117-123, (1896).
   do   Quinn, Garden and Field, 22, No. 1, pp. 24, 25, (1896).
   do   King, Can. Ent., p. 228, (18—).
   do   Perkins, Vt. Agr. Exp. Sta., Bull. No. 60, pp. 9-12, (1897).
   do   Starnes, Ga. Sta., Bull. 36, (1897).
   do   Panton, Ont. Agr. Coll. and Exp. Farm, Bull. 105, p. 15, (1897).
   do   Barrows and Pettit, Mich. Sta., Bull. 160, pp. 339-436, (1897).
   do   Lowe, N. Y. Agr. Exp. Sta., Bull. No. 136, p. 578, (1897).
   do   Fletcher, Exp. Farms, pp. 187-230, (1897).
   do   Lea, Producers' Gaz. and Settlers' Rec., 5, No. 6, pp. 465-483, (1897).
   do   Gillette, Colo. Agr. Coll. Exp. Sta., Bull. No. 47, p. 13, (1898).
   do   Parrott, Trans. Kan. St. Hort. Soc., Bull. 23, pp. 106-109, (1898).
   do   Hunter, Bull. Dept. Ent., p. 25, (Jan. 1898).
   do   Gillorders, Reprint Trans. Ann. Manchester Micr. Soc., p. 7, (1898).*
   do   Chambliss, Tenn. Sta. Bull., vol. X, No. 4, pp. 141-151, (1898).
   do   Leonardi, Annali di Agr., p. 133, (1898).

Mytilaspis pomorum Newstead, Ent., XXXI, p. 98, (1898).
do   Brick, Station fur Pflanzenschutz zu Hamburg, 1, p. 34, (1898-'99).
do   Bruner, Ins. Enemies Apple Tree, p. 146, (1899).
do   Smith, N. J. Agr. Coll. Exp. Sta., Bull. No. 140, p. 7, (18—).
do   Newell, Iowa Sta., Bull. 43, pp. 159, 160, (1899).*
do   Fernald, Com. Pa. Dept. Agr., Bull. No. 43, p. 16, (1899).
do   Johnson, U. S. Dept. Agr., Div. Ent., Bull. 20, n. s., p. 66, (1899).
do   Marchal, Bull. Soc. D'Arc. de France, p. 11, (1899).
do   Troop, Ind. Agr. Exp. Sta., Bull. No. 78, vol. X, (1899).
do   Reh, Mitt. Naturhist. Mus. Hamburg, XVI, pp. 125-141, (1899).*
do   Leonardi, Lab. di Ent. Agr. in Portici, p. 5, (1899).
do   Hunter, K. U. Quart., vol. VIII, No. 1, p. 14, (1899).
do   Proc. Acad. Nat. Sci. Phila., p. 275 (1899).*
do   Fletcher, Rept. Ent. and Bot., p. 17, (1900).
do   Reh, Jahrb. Hamb. Wiss. Anst., XVII, pp. 3-5, (1900).* Zeitschrift fur Ent., V, p. 161, (1900).*
do   Hempel, Rev. Mus. Paulista, IV, pp. 512, 513, (1900).*
do   Lochhead, Bull. Dept. Agr., p. 40, (1900).
do   Frank and Kruger, Schildlausbuch, pp. 90-98, (1900).*
do   Harvey and Munso, Me. Sta. Bull. 56, pp. 105-144, (——).

Mytilaspis citricola Packard.

Mytilaspis citricola Packard, Guide to Study Insects, 2d ed., p. 527, (1870).
do   Comstock, U. S. Dept. Agr. Rept. for 1880, p. 321.
do   Hubbard, Rept. U. S. Dept. Agr., Div. Ent., pp. 107-109, (1881)*; pp. 157-159, (1883).*
do   Comstock, 2d Rept. Ent. Corn. Univ. Exp. Sta., p. 117, (1883).

Mytilaspis flavcocus Targioni, Annali di Agricultura, p. 392, (1884).*
Mytilaspis citricola Voyle, Bull. 4 (O. S.), Div. Ent. U. S. Dept. Agr., pp. 70-73, (1884).*
do   Maskell, Ent. Mo. Mag., XXVIII, p. 70, (1892).*
do   Cockerell, Ent. Mo. Mag., XXIV, p. 155, (1893).*
do   Craw, Cal. St. Bd. Hort. Rept., pp. 90-96, (1894-'94).
do   Cockerell, Journ. Trinidad Field Nat. Club, p. 306, (1894).*
do   Smith, Ent. News, V, pp. 184, 185, (1894).*
do   Green, Coccidæ Ceylon, pt. I, pp. 78-80, (1896).*
do   Marlatt, Dept. Agr., Year book, pp. 217-236, (1896).
do   Cockerell, Bull. Dept. Bot. Jamaica, IV, n. s., pp. 108, 109, No. 65, (1897).*
do   Froggatt, Agr. Gaz., N. S. Wales, 9, No. 10, pp. 1210-1221, (1898).
do   Maskell, Ent. Mo. Mag., XXVII, p. 70, (1898).
do   Producers' Gaz. and Settlors' Rec. (West. Aus.), 5, No. 6, pp. 465-483, (1898).
do   Frank and Kruger, Schildlausbuch, p. 99, (1900).*
do   Hempel, Rev. Mus. Paulista, IV, pp. 513, 514, (1900).*
do   Reh, Zeits. fur. Ent., V, p. 162, (1900).*
do   Cockerell, Rev. Mus. Paulista, IV, p. 363, (1900).*
do   Morgan, La. Sta. Bull. 28, 2d ser., pp. 982-1005, (——). Sp. Bull., pp. 51-110 (——).

Diaspis snowii.

Diaspis snowii Hunter, K. U. Quart., vol. 8, No. 1, p. 14, (1899).

Lecanium aurantiacum.

Lecanium maclurœ Hunter, K. U. Quart., vol. 8, No. 2, p. 67, (1899).
Lecanium aurantiacum Hunter, K. U. Quart., vol. 9, No. 2, p. 107, (1900).

Lecanium canadense.

Lecanium caryœ, var. canadense Ckll., Can. Ent., vol. 27, p. 253, (1895).
Lecanium canadense Ckll., Can. Ent., vol. 30, p. 294, (1898).

    do   Cockerell and Parrott, Industrialist, p. 232, (1899).
    do   Hunter, K. U. Quart., vol. 8, No. 2, p. 68, (1899).
    do   King, Can. Ent., vol. XXXI, p. 252, (Sept. 1899).
    do   Hunter, K. U. Quart., IX, ser. A., No. 2, p. 107, (Apr. 1900).
    do   King, Psyche, IX, p. 117, Oct. 1900.

Lecanium kansasense.

Lecanium kansasense Hunter, K. U. Quart., vol. 8, No. 2, p. 69, (1899).

Lecanium cockerelli Hunter.

Lecanium cockerelli Hunter, K. U. Quart., vol. VIII, No. 2, p. 70, (1899).
                        K. U. Quart., vol. IX, No. 2, p. 107, (1900).
    do   King, Can. Ent., vol. —, p. 252.
                    Psyche, IX, p. 117, Oct. 1900.

Lecanium armeniacum.

Lecanium armeniacum Craw, Cal. St. Bd. Hort., Div. Ent., pp. 12, 13, 1891.
    do   Cockerell, Can. Ent., vol. 26, No. 2, p. 35, (1894).
    do   Felt, Bull. N. Y. St. Mus., vol. V, No. 23, p. 240, (1898).
    do   Webster, Can. Ent., p. 81, (Apr. 1898).
    do   Cockerell and Parrott, Industrialist, p. 233, (Apr. 1899).
    do   Cockerell, Journ. N. Y. Ent. Soc., VII, p. 257, Dec. 1899.*
    do   Maskell, Trans. N. Z. Inst., XXVII, p. 59, 1894, (1895)*.
    do   King, Psyche, IX, p. 117, Oct. 1900.
    do   Hunter, K. U. Quart., VIII, ser. A, No. 2, pp. 71–75, pl. XV, fig. 4, Apr. 1899.

Lecanium hesperidum Linn.

Coccus hesperidum Linn., Syst. Nat., II, 739, (1735).
    do   Faun. Suec., 1015, (1746).
    do   Frish, Ins., 12, 13, (1736).
    do   Sulzer, Ins., p. 1091, tab. 12, fig. 8, (1761).
    do   Fabricius, Syst. Entom., p. 743, (1775).
    do   Modeer, Act. Gothenb., 1, 19, 8, (1778).
    do   Schaeffer, Eelem. Entom., tab. 48, (1766).
    do   Schrank, Enum. Insect Austriæ, pp. 295, 583, (1781).
    do   Gmelin, Syst. Nat., 2215, (1788).
    do   De Villers, Linnœi Entomologia (Syst. Nat.?), p. 558, (1789).
    do   Olivier, Encycl. Method, VI, 93, (1791).
    do   Fabricius, Entom. Syst., IV, 244, et Syst. Ryng., 306.
    do   Schaeffer, Icones Insect, tab. CXI, fig. 2, (1801).

Coccus hesperidum Gene, Insetti Nocivi, etc., p. 113, tab. II, fig. 12, (1827).
   do   Fonscolombe, Ann. Soc. Ent. Fr., 3, vol. 208, (1834).
   do   Burmeister, Handb. der Entomologie, p. 69, (1835).
Calymnatus hesperidum Costa (O. G.), Nueve Observ., tab. 1, fig. 1, (1835).
Calypticus lævis Costa (O. G.), Faun. Ins. Nap. Gallins, 8, 1, (1837).
Lecanium hesperidum Blanchard, Hist. Nat. Ins., (1840).
Calypticus hesperidum Lubbock, Proc. Roy. Soc., IX, 480, (1858).
                    Proc. Roy. Soc., IX, 480, and Ann. Nat.
                      Hist., III, 306, (1859).
   do   Beck, Trans. Micr. Soc. London, n. s., 47, (1861).
   do   Boisduval, Entom. Hort., 331, (1867).
   do   Targioni-Tozzetti, Catal., 37, 5, (1868).
Lecanium hesperidum Signoret, Ann. Soc. Ent. Fr. p. 830, 856, (1868).
   do   Targioni-Tozzetti, Intr. Seconda Mem. Coccin., p. 37, (1868).
   do   Signoret, Ann. Soc. Ent. Fr., p. 399, (1873).
   do   Maskell, Trans. N. Z. Inst., XI, pp. 205, 206, pl. VI, fig. 12, 1878, (1879).*
   do   Ashmead, Orange Insects, pp. 32, 33, fig, 10; pl. I, fig. 14, 1880.*
   do   Comstock, Rept. Ent., in Rept. Comr. Agr., 1880, p. 335.
   do   Targioni-Tozzetti, Ann. di Agr. (R. Minist. Agr.), p. 142, (1881).
   do   Comstock, 2d Rept. Dept. Ent. in Corn. U. Exp. Sta., p. 134, (1883).
   do   Hubbard, Ins. aff. Orange., U. S. Dept. Ent., p. 48, (1885).
   do   Douglas, Ent. Mo. Mag., XXII, p. 159, Dec. 1885.*
   do   M. Cooke, Treat. Ins. Inj. to Fruit and Fruit-trees, Sacramento, p. 36, (1886).
   do   Atkinson, Journ. Asiat. Soc. Bengal, LV, pt. II, No. 3, pp. 281, 282, 1886.*
   do   Maskell, Sca. Ins. N. Z., p. 80, (1887).
   do   Penzig, Studi Botanici sagli a grumi e sulle piante atfin (Ann. Min. Agr.), p. 521, (1887).
   do   Douglas, Ent. Mo. Mag., XXIV, pp. 25-27, July, 1887.*
   do   Moniez, Bull. Soc. Zool. Fr., XII, pp. 150-152, 1887.*
   do   Douglas, Ent. Mo. Mag., XXVII, pp. 244, 245, pl. 2, fig. 1, Sept. 1891.*
   do   Wassilieff, Trav. Soc. Varsavie, I1, No. 6, pp. 10-12, 1891.*
   do   Berlese, Rivista Pat. Veg., I, p. 61 (58-70), figs. 4-6, 1892.*
   do   Coquillett, U. S. Dept. Agr., Div. Ent., Bull. No. 26, p. 26, (1892).
   do   Maskell, Trans. N. Z. Inst., XXV, pp. 218, 219, 1892, (1893).*
   do   Cockerell, Trans. Am. Ent. Soc., April, 1893, p. 49.
   do   Maskell, Ent. Mo. Mag., XXIX, pp. 103, 104, May, 1893.*
   do   G. del Guercio, Staz. Sper. Agr. Ital., 24, pp. 573-592, (1893).
   do   Cockerell, Ann. and Mag. Nat. Hist., ser. 6, p. 52, (1893).

Lecanium hesperidum Koebele, Dept. Agr., Div. Ent., Bull. 32, pp. 33-36, (1893).
do   Cockerell, Trans. Am. Ent. Soc., vol. XX, p. 49, 1893.*
                 Bull. Bot. Dept. Jamaica, Feb. 1894, pp. 18, 173, No. 13.*
do   Douglas, Ent. Mo. Mag., XXX, p. 29, Feb. 1894.*
do   Cockerell, Ent. News, p. 210, (Sept. 1894).
do   Berlese, Le Cocciniglie Italicne, pt. II, p. 113, (1894).
do   Smith, Rept. Ent., N. J. Agr. Coll. Exp. Sta., p. 501, (1894).
do   Cockerell, Act. Soc. Sci. Chili, V, p. 24, 1895.*
do   Gilletto, Colo. Agr. Exp. Sta., Bull. No. 31, Tech. Ser., No. 1, p. 127, (1895).
do   Townscnd, U. S. Dept. Agr., Div. Ent., Tech. Ser., No. 4, p. 11, (1896).
do   Cockerell, Amcr. Nat., July, 1897, p. 590.*
do   Osborn, Proc. Iowa Acad. Sci., V, p. 226, 1897, (1898).*
do   Producers' Gaz. and Settlers' Rec. (W. Aus.), V, No. 6, pp. 465-483, (1898).
do   Hunter, K. U. Quart., vol. VIII, No. 2, p. 75, (1899).
do   King, Can. Ent., vol. XXXI, p. 140, (June, 1899).
do   Gossard, Fla. Exp. Sta., Bull. No. 51, p. 113, (Jan. 1900).

Lecanium coffeæ.

Lecanium coffeæ Cockerell, Bull. Bot. Dept. Jamaica, vol. 1, P. 5, p. 71, (May, 1894).
do   Cockerell and Parrott, Cont. to Knowledge of Coc., XXX, Industrialist, vol. —, No. —, p. 164, (1899).
do   Cockerell, Nat. Sci. Phila., p. 270, (1899).
do   Hunter, K. U. Quart., vol. 8, No. 2, p. 75, (1899).
do   King, Can. Ent., vol. XXXI, p. 140, (June, 1899).
do   Hempel, Rev. Mus. Paulista, IV, p. 426, (1900).*
do   Cockerell, Rev. Mus. Paulista, IV, p. 363, (1900).*
do   Watt, Pests and Blights of the Tea Plant, p. 330, 1898.*
do   Atkinson, Journ. Asiat. Soc. Bengal, LV, pt. II, No. 3, pp. 282-234, 1886.*
do   Maskell, Ent. Mo. Mag., XXVIII, p. 71, Mar. 1892.*

Lecanium oleæ Bernard.

Chermes oleæ Bernard, Mem. d' Hist. Nat. Acad., 108, pl. 2, fig. 25, (1782).
Coccus oleæ Olivier, Encycl. Method., 95, 6, (1791).
do   Giovene, Mem. Soc. Italie, vol. XIV, p. 128, (1809).
do   Fonscolombe, Soc. Ent. Fr., p. 206, (1834).
do   (ex. p.) Costa (A.), Insctti dell' Olivo, etc., p. 111, (1840).
Lecanium oleæ Walker, List of Hompt. in the Coll. of Brit. Mus., P. IV, p. 1070, (1852).
do   Boisduval, Entom. Hortic., 319, fig. 38, (1867).
do   Targioni-Tozzetti, Catalogo, etc., pp. 39-22, (1868).
do   Signoret, Essai sur les Cochen, p. 440, (1873).
do   Goureau, Ann. Soc. Fr., 2d Sèrie, 11.
do   Costa (A.), Degl. Insetti che Attacana l'Alboro ed il Frutto dell' Olivo, p. 122, pl. IV, A, figs. 11, 12, 1877.*

Lecanium oleœ Comstock, Rept. Ent., in Rept. Com. Agr. 1880, p. 336.
do   Signoret, Treatise on Ins. Inj. to Frt. and Frt. Trees, by M.
     Cooke, Sacr., 1881, p. 35.
do   Chapin, 1st Kept. St. Bd. Hort. Com. Cal., pp. 65-68, 1882.*
     Pac. Rural Press, Sept. 28-Nov. 24, 1882.*
do   Hubbard, Ins. aff. Orange, U. S. Dept. Ent. 1885, p. 53.
do   Riley, Bull. 12 (O. S.), Div. Ent. U. S. D. A., pp. 34-36, 1885,
     (1886).*
do   Maskell, Sca. Ins. New Zealand, p. 82, (1887).
do   Penzig, Studi Botan. Agrumi, etc. (Ann. Min. Agr.), p. 527, (1887).
do   Douglas, Ent. Mo. Mag., XXVII, pp. 307, 308, Nov. 1891.*
do   Coquillett, U. S. Dept. Agr. Div. Ent., Bull. No. 26, p. 28-33, (1892).
do   Berlese, Riv. Pat. Veg., I, p. 62 (58-70), 1892.*
     Riv. Pat. Veg., III, Nos. 1-8, 1893.*
do   Cockerell, Trans. An. Soc. Ent., p. 55, (Apr. 1893).
do   Craw, Cal. St. Bd. Hort. Rept., pp. 90-96, (1893-'94).
do   Cockerell, Can. Ent., vol. ——, p. 44, (——).
· do   Berlese, Le Cocciniglie Italiane, part II, p. 120, (1894).
do   Cockerell, Bull. Bot. Dept. Jamaica, I, p. V, p. 72, (May, 1894).
     Journ. Trin. Field Nat. Club, I, p. 307, No. 7, Feb. 1894.*
     Journ. Trin. Field Nat. Club, I, No. 12, p. 307, (1894).
     Amer. Nat., XXIX, p. 727, Aug. 1895.*
do   Gillette, Colo. Agr. Exp. Sta., Bull. No. 31, Tech. Ser., No. 1,
     p. 127, (1895).
do   Toumey, Ariz. Sta., Bull. 14, pp. 29-56, (1895).
do   Saccardo, Riv. Pat. Veg., IV, pp. 48, 49, 1895, (1895-'96).*
do   Ckll., Appx. Bull. Misc. Information (Trinidad), II, pp. III and
     IV, No. 23, Apr. 1896.*
do   Marlatt, Year-book Dept. Agr., p. 220, (1896).
do   Ckll., Cal. Fruit Grower, XX, May 8, 1897, p. 4.*
do   Maskell, Tr. N. Zeal. Inst., XXIX, pp. 309, 310, 1896, (1897).*
do   Green, Ent. Mo. Mag., XXXIII, p. 72, Mar. 1897.*
do   Lea, Producers' Gaz. and Settlers' Rec. (W. Aue.), 5, No. 6, pp.
     465-483, (1898).
do   Froggart, Agr. Gaz. N. S. Walee, 9, No. 10, pp. 1216-1221, (1898).
do   Leonardi, Lab. di. Ent. Agr. in Portici, p. 7, (1899).
do   Hunter, K. U. Quart. vol. VIII, No. 2, p. 75, (1899).
do   Guercio, Atti R. Accad. Econ. Agr. Georg. Firenze, 4 ser., 22,
     No. 1, pp. 50-76, (1899).
· do   Morgan, La. Sta., Sp. Bull. —, pp. 51-110, (1899).
do   Cockerell and Parrott, Industrialist, Mar. 1899, p. 163.
do   Fuller, Tr. Ent. Soc. London, 1899, part IV, pp. 459, 460.*
do   Gossard, Fla. Agr. Exp. Sta., Bull. No. 51, p. 115, (Jan. 1900).

Lecaniodiaspis (?) parrotti.

Lecaniodiaspis parrotti Hunter, K. U. Quart., vol. 8, No. 2, p. 76, (1899).

Lecaniodiaspis celtidis pruinosus.

Lecaniodiaspis celtidis pruinosus Hunter, K. U. Quart., vol. 8, No. 2, p. 77, (1899).

Chionaspis ortholobis.

Chionaspis ortholobis Comstock, Rept. Ent., in Rept. Com. Agr., p. 317, (1880).
2d Rept. Dept. Ent., Corn. U. Exp. Sta., p.
105, (1883.)
do   Packard, 5th Rept. U. S. Ent. Com., p. 594, (1890).
do   Cockerell, Can. Ent., vol. XXVI, p. 189, (1894).
do   Howard, Ins. Life, vol. VI, p. 328, (1894).
do   Gillette and Baker, Hemipt. of Colo., p. 129, (1895).
do   Cockerell, Bull. 24, N. M. Agr. Exp. Sta., p. 38, (1897).
do   Osborn, Contr. from Dept. Zool. and Ent., Iowa Agr.
Coll., No. 3, p. 5, (1898).
do   Cooley, Hatch Exp. Sta., Mass. Agr. Coll., Spec. Bull.,
pp. 17, 18, (1899).
do   Hunter, K. U. Quart., vol. 9, No. 2, p. 101, (1899).
do   Newell, Bull. Iowa Agr. Sta. No. 43, pp. 154, 155, (1899).
do   King, Psyche, IX, p. 117, Oct. 1900.

Chionaspis salicis-nigræ.

Aspidiotus salicis-nigræ Walsh, 1st Rept. Nox. Ins. Ill., p. 39, (1867).
Mytilaspis salicis Le Baron, Trans. Ill. Hort. Soc., App., p. 140, (1871).
2d Rept. St. Ent. Ill., p. 140, (1872).
Chionaspis salicis Comstock, Rept. U. S. Dept. Agr., p. 320, (1881).
do   Osborn, Trans. Iowa St. Hort. Soc., vol. XVII, p. 214, (1882).
do   Comstock, 2d Rept. Ent., Corn. U. Exp. Sta., p. 106, (1883).
Intro. Ent., part I, p. 151, (1888).
do   Packard, 5th Rept. U. S. Ent. Com., p. 593, (1890).
Mytilaspis salicis Forbes, 17th Rept. Nox. and Benif. Ins. Ill., App., p. 23, (1891).
Chionaspis salicis Lugger, Bull. 43, Minn. Agr. Exp. Sta., p. 224, (1895).
do   salicis-nigræ Gillette and Baker, Bull. 31, Colo. Exp. Sta., p. 129,
May, 1895.*
Chionaspis salicis Lugger, 1st Ann. Rept. Ent. Minn., p. 128, (1895).
do   Osborn, Proc. Iowa Acad. Sci., vol. V, p. 224, (1898).
Contr. from Dept. Zool. and Ent. Iowa Agr. Coll.,
No. 3, p. 4, (1898).
Chionaspis ortholobis bruncri Cockerell, Can. Ent., vol. XXX, p. 135, (1898).
Chionaspis salicis-nigræ Cooley, Hatch Exp. Sta. Mass., Spec. Bull., pp. 19-22,
(Aug., 1899).
do   Hunter, K. U. Quart., vol. 9, No. 2, p. 101, (1899).

Chionaspis americana.

Chionaspis americana (Johnson MS.), Howard, Bull. U. S. Dept. Agr., Div.
Ent., Tech. Ser., No. 1, p. 44, (1895).
do   Lugger, Minn. Sta. Bull. 43, p. —, (1895).
do   Johnson, Ent. News, vol. VII, p. 150, (1896).
do   Johnson, Bull. Ill. St. Lab. Nat. Hist., vol. IV, p. 390,
(1896).
do   Johnson, Bull. 6, n. s., Div. Ent. U. S. Dept. Agr.,
p. 177, 1896.
do   Lugger, 1st Ann. Rept. Ent. Minn., p. 129, (1896).
do   Cooley, Hatch Exp. Sta., Mass. Agr. Coll., Spec. Bull.,
pp. 41-43, (1899).

Chionaspis americana Hunter, K. U. Quart., vol. 9, No. 2, p. 102, (1899).
 do   Newell, Iowa Sta., Bull. 43, p. 152, (1899).
 do   King, Psyche, vol. IX, p. 117, Oct. 1900.

Chionaspis platani.

Chionaspis platani Cooley, Hatch Exp. Sta., Mass. Agr. Coll., Spec. Bull.,
     p. 36, (Aug. 1899).
 do   Hunter, K. U. Quart., vol. 9, No. 2, p. 102, (1899).

Chionaspis pinifoliæ.

Aspidiotus pinifoliæ Fitch, Tr. N. Y. St. Agr. Soc., vol. XV, p. 488, (1856).
     2d. Rept. Nox. Benif. Ins. N. Y., (1856).
     Tr. N. Y. St. Agr. Soc., vol. XVII, p. 741, (1858).
     4th Rept. Nox. Benif. Ins. N. Y., (1858).
 do   Walsh, Pract. Ent., vol. I, p. 90, (1866).
Mytilaspis pinifoliæ Le Baron, 1st Ann. Rept. Nox. Ins. Ill., p. 83, (1871).
     2d Ann. Rept. Nox. Ins. Ill., p. 161, (1872).
 do   Riley, 5th Ann. Rept. Nox. Benif. Ins. Mo., p. 97, (1873).
 do   Bessey, Rept. Iowa St. Agr. Soc. 1874, p. 232, (1875).
Chionaspis pinifoliæ, Comstock. Ann. Rept. U. S. Dept. Agr., 1880, p. 318,
     (1881).
Mytilaspis pinifoliæ Packard, Ins. Inj. to Forest and Shade Trees, Bull. 7, U.
     S. Ent. Com., p. 218, 1881).
Chionaspis pinifoliæ Riley, Am. Nat., vol. XVI, p 514, (1882).
 do   Comstock, 2d Rept. Dept. Ent. in Corn. U. Exp. Sta ,
     1883, p. 105.
 do   Saunders, Rept. Ent. Soc. Ont. 1883, p. 52, (1884).
 do   Lintner, 2d Ann. Rept. Inj. Ins. N. Y., p. 184, (1885).
     5th Ann. Rept. Inj. Ins. N. Y., p. 266, (1889).
Mytilaspis pinifoliæ Packard, 5th Rept. U. S. Ent. Com., p. 805, (1890).
Chionaspis pinifoliæ Lintner, 7th Ann. Rept. Inj. Ins. N. Y., p. 384, (1891).
     9th Ann. Rept. Inj. Ins. N. Y., p. 376, (1893).
 do   Howard, Bull. U. S. Dept. Agr., Div. Ent., Tech. Ser.
     No. 1, pp. 13, 22, 52, (1895).
 do   Lintner, Bull. N. Y. St. Mus., 3, No. 13, pp. 267–305, (1895).
 do   Comstock, Man. Study Ins., p. 174, (1895).
 do   Gillette and Baker, List Hemipt. Colo., Bull. Colo. Agr.
     Exp. Sta. No. 31, p. 129, (1895).
 do   Ckll., Am. Nat., XXIX, pp. 730, 731, Aug. 1895.
 do   Lintner, 10th Ann. Rept. Inj. Ins. N. Y., p. 518, (1895).
 do   Webster, Ind. Hort. Rept. 1896, p. 16.
 do   Johnson, Dept. Agr., Div. Ent., pp. 75–78, (1896).
 do   Lintner, 11th Ann. Rept. Inj. Ins. N. Y., p. 203, (1896).
 do   Cockerell, Bull. N. M. Agr. Exp. Sta. No. 24, p. 38, (1896).
 do   Barrows and Pettit, Mich. Sta., Bull. 60, pp. 339–436,
     (1897).
 do   Osborn, Proc. Iowa Acad. Sci., vol. V, p. 224, (1898).
 do   Gillette, Bull. Colo. Agr. Exp. Sta. No. 47, p. 36, (1898).
 do   Pettit, Bull. Mich. Agr. Exp. Sta. No. 160, p. 415, (1898).
 do   Cooley, Hatch Exp. Sta., Mass. Agr. Coll., Spec. Bull.,
     pp. 30–31, (Aug. 1899).
 do   Hunter, K. U. Quart., vol. IX, No. 2, p. 101, (1899).
 do   King, Can. Ent., p. 252.
 do   Newell, Iowa Sta., Bull. 43, pp. 157, 158, (1899).

Chionaspis furfura.

—— ? —— Harris, Rep. Ins. of Maes. Inj. Veg., p. 202, (1841).
Aspidiotus furfurus Fitch, Trane. N. Y. St. Agr. Soc., vol. XVI, p. 352, (1856).
Aspidiotus cerasi Fitch, Trans. N. Y. St. Agr. Soc., vol. XVI, p. 368, (1856).
Aspidiotus furfurus Fitch, 3d Rept. Nox. and other Ins., p. 352, (1857).
Aspidiotus cerasi Fitch, 3d Rept. Nox. and other Ins., p. 368, (1857).
—— ? —— Harris, Treat. Ins. Inj. Veg. (3d ed.), p. 254, (1862).
Coccus ? harrisii Walsh, Pract. Ent., vol. II, p. 31, (1866).
Aspidiotus harrisii Walsh, Pract. Ent., vol. II, p. 119, (1867).
　　　　　　1st Rept. Nox. Ins. Ill., pp. 36–53, (1868).
do　　　Riley, 1st Ann. Rept. Ins. Mo., p. 7, (1869).
　　　　　　Am. Ent., vol. II, pp. 110–181, (1870).
do　　　2d Ann. Rept. Ins. Mo., p. 9, (1870).
do　　　Bethune, Rept. Ent. Soc. Ont., I, p. 303, (1870).
do　　　Glover, Ann. Rept. U. S. Dept. Agr., 1870, p. 88, (1871).
do　　　Beseey, Rept. Iowa St. Agr. Soc. 1874, p. 232, (1875).
Diaspis harrisii Signoret, Ann. Soc. Ent. Fr., ser. 4, vol. XVI, p. 604, (1876).
Aspidiotus harrisii Thomas, 7th Rept. Ins. Ill., p. 108, (1878).
Chionaspis furfurus Comstock, Ann. Rept. U. S. Dept. Agr., 1880, p. 315, (1881).
Diaspis harrisii Riley, Am. Nat., vol. XV, p. 487, (1881).
Chionaspis furfurus Lintner, 1st Ann. Rept. Inj. Ins. N. Y., p. 331, (1882).
Aspidiotus harrisii Packard, Guide Study Ins., p. 530, (1883).
Chionaspis furfurus Osborn, Tr. Iowa St. Hort. Soc., vol. XVII, p. 211, (1883).
do　　　Hagen, Can. Ent., XVI, pp. 161–163, (1884).*
do　　　Comstock, Intro. to Ent., part I, p. 151, (1888).
do　　　Lintner, 4th Ann. Rept. Inj. Ins. N. Y., p. 208, (1888).
do　　　Tryon, Rept. Ins. and Fung. Pests, No. 1, p. 89, (1889).
do　　　Riley-Howard, Ins. Life, vol. I, p. 324, (1889).
do　　　Lintner, 5th Ann. Rept. Inj. Ins. N. Y., pp. 300–326, (1889).
Aspidiotus cerasi Saunders, Ins. Inj. to Fruits, p. 204, (1889).
"A species of *Coccus*" Downing, Fruits and Fruit-trees of America, p. 66, (1890).
Chionaspis furfurus Riley-Howard, Ins. Life, vol. III, p. 4, (1890).
do　　　Weed (C. M.), Bull. Ohio Agr. Exp. Sta., vol. III, No. 4,
　　　　　　p. 128, (1890).
do　　　Packard, 5th Rept. U. S. Ent. Com., p. 537, (1890).
do　　　Weed (C. M.), Ins. Insecticides, p. 66, (1891).
do　　　Townsend, N. Mex. Sta., Bull. No. 3, p. 19, (1891).
do　　　Gillette, Ins. Life, vol. III, p. 259, (1891).
do　　　Weed (C. M.), Ann. Rept. Colum. Hort. Soc. 1890, p. 16,
　　　　　　(1891).
do　　　Troop, Trans. Ind. Hort. Soc. 1891, p. 75, (1892).
do　　　Morgan, Ent. Mo. Mag., vol. XXIX, p. 16, (1892).
do　　　Webster, Bull. Ohio Agr. Exp. Sta. No. 45, p. 208, (1892).
do　　　Lintner, 8th Ann. Rept. Inj. Ins. N. Y., pp. 293–299, (1893).
do　　　Osborn, Rept. Iowa St. Hort. Soc., XXVII, p. 122, (1893).
do　　　Marlatt, Ins. Life, vol. VII, p. 120, (1894).
do　　　Smith, Ann. Rept. N. J. Agr. Exp. Sta., 1894, p. 496, (1894).
do　　　Bruner, Ann. Rept. Neb. St. Hort. Soc., 1894, p. 175, (1894).
do　　　Howard, Ins. Life, vol. VII, p. 5, (1894).
do　　　Smith, Ins. Life, vol. VII, p. 186, (1894).
do　　　Howard, Can. Ent., vol. XXVI, p. 354, (1894).

Chionaspis furfurus Lintner, Bull. N. Y. St. Mus. 3, No. 13, pp. 266–305, (1895).
do   Howard, Year-book U. S. Dept. Agr. 1884, p. 259, (1895).
do   Garman, Ky. Sta. Rept., pp. 32-57, (1895).
do   Lintner, 10th Ann. Rept. Inj. Ins. N. Y., p. 518, (1895).
do   Webster, Ind. Hort. Rept., p. 7, (1896).
do   Fletcher, Ann. Rept. Can. Exp. Farm, 1895, p. 148, (1896).
do   Hopkins, Dept. Agr., Div. Ent., Bull. No. 6, n. s., pp. 71-74, (1896).
do   Howard, Trans. Mass. Hort. Soc. 1896, p. 89, (1896).
do   Johnson, Dept. Agr., Div. Ent., Bull. No. 6, n. s., pp. 75-78, (1896).
do   Lintner, 11th Ann. Rept. Inj. Ins. N. Y., pp. 202, 271, 288, (1896).
do   Garman, 8th Ann. Rept. Ky. Agr. Exp. Sta., p. 37, (1896).
do   Smith, Econ. Ent., p. 119, (1896).
do   Coons, Rept. Sec. Bd. Agr. Conn. 1896, p. 16, (1896).
do   Starnes, Bull. Ga. Agr. Exp. Sta. No. 36, p. 27, (1897).
do   Lowe, Bull. N. Y. Agr. Exp. Sta., p. 582, (1897).
do   Starnes, Ga. Sta., Bull. 36, (1897).
do   Lintner, 12th Ann. Rep. Inj. Ins. N. Y., p. 348, (1897).
do   Webster, Bull. Ohio Agr. Exp. Sta. No. 81, p. 210, (1897).
do   Lintner, Country Gentleman, July 8, 1897.
do   Osborn, Proc. Iowa Acad. Sci., vol. V, p. 224, (1898).
do   Parrott, Trans. Ks. St. Hort. Soc., vol. 23, pp. 106-109, (1898).
do   Gillette, Bull. Colo. Agr. Exp. Sta. No. 47, p. 12, (1898).
do   Barrows and Pettit, Mich. Sta., Bull. No. 160, p. 415, (1898).
do   Kirkland, Mass. Crop Rep., June, 1898, p. 28, (1898).
do   Pettit, Bull. Mich. Agr. Exp. Sta., p. 415, (1898).
do   Osborn, Contr. from Dept. Zool. and Ent., Iowa Agr. Coll., No. 3, p. 4, (1898).
do   Bruner, Ins. Enemies Apple Tree, p. 147, (1899).
do   var. *fulvus* King, Psyche, vol. VIII, p. 334, (1899).
               King, Can. Ent., p. 251, (18—).
do   Cooley, Hatch Exp. Sta., Mass. Agr. Coll., p. 23, (Aug. 10, 1899).
do   Fernald, Com. Pa. Dept. Agr., Bull. No. 43, p. 18, (1899).
do   Troop, Ind. Agr. Exp. Sta., Bull. 78, vol. X, (1899).
do   Smith, N. J. Agr. Coll. Exp. Sta., Bull. No. 140, p. 7, (1899).
do   Johnson, U. S. Dept. Agr., Div. Ent., Bull. 20, n. s., pp. 62-68, (1899).
do   King, Can. Ent., p. 251, (18—).
do   Reh, Mitt. Naturh. Mus. Hamburg, XVI, pp. 125-141, (Mar. 1899).
do   Newell, Iowa Sta., Bull. 43, pp. 150-152, (1899).
do   Felt, Bull. N. Y. St. Mus., vol. 6, No. 31, p. 578, (1900).
do   Lochhead, Ont. Dept. of Agr., p. 42, (Mar. 1900).
do   Chambliss, Tenn. Sta. Bull., vol. X, No. 4, pp. 141-151, (18—).
do   Frank and Krüger, Schildlausbuch, pp. 100, 101, 1900.
—   do   Cockerell, Science, XI, n. s., p. 671, Apr. 27, 1900.

Pulvinaria innumerabilis.

Coccus innumerabilis Rathvon, Pa. Farm. Journ., vol. IV., pp. 256–258, (1851).

Lecanium accricorticis Fitch, Trans. N. Y. St. Agr. Soc., 1859, vol. XIX, pp. 775, 776, (1860).

Coccus aceris Leidy, Rept. to Councils Phila. on Ins. Inj. to Shade-trees, 1862.

Lecanium acericola Walsh and Riley, Amer. Ent., vol. I, pp. 14, 15, (1868). (A wrong determination.)

Lecanium macluræ Walsh and Riley, Amer. Ent. vol. I, p. 14, (1868).

Lecanium accrella Rathvon, Lancaster Farmer, vol. VIII, pp. 101, 102, (1876).

Pulvinaria innumerabilis Comstock, Rept. Ent., in Rept. Com. Agr. 1880, p. 334, (1880).

	2d Rept. Dept. Ent. in Corn. Univ. Exp. Sta., p. 137, (1883).
do	Mann, Psyche, IV, p. 224, (1884).
do	Saunders and Mundt, Can. Ent., vol. 16, pp. 141, 143, 210, 211, 240, (1884).
do	Riley, Rept. U. S. Ent., pp. 350–355, (1885).*
do	Weed, Ohio Sta. Bull., vol. III, No. 11, 2d ser., p. 72, (1890).*
do	Garman, Ky. Sta., Bull. No. 39, p. 11, (1892).
	Ky. Sta., Bull. No. 47, pp. 3–53, (1893).
do	Cockerell, Science, XXII, p. 78, (1893).*
do	Smith, Rept. Ent., N. J. Agr. Exp. Sta., pp. 505–509, (1894).
do	Lugger, Minn. Sta., Bull. 43, pp. 99–252, (1895).
do	Lintner, Bull. N. Y. St. Mus., 3, No. 13, pp. 267–305, (1895).
do	Johnson, Pa. Dept. Agr. Rept., pp. 345–373, (1896).
do	Hubbard, Proc. Ent. Soc. Washington, vol. III, p. 319, (1896).
do	Cockerell, Ent., vol. XXX, No. 404, pp. 12–14, (1897).
do	Gillette, Colo. Sta. Rept., pp. 55–61, (1897).
do	Hunter, Bull. Dept. Ent. Univ. Kan., pp. 25–27, (1898).
do	Felt, Bull. N. Y. St. Mus., vol. V, No. 23, p. 239, (1898).
do	Hopkins, Dept. Agr., Div. Ent., Bull. 17, n. s., pp. 44–49, (1898).
do	Felt, Ext. from 4th Ann. Rept. Com. on Fishes, etc., p. 29, (1898).
do	Gillette, Colo. Agr. St. Exp. Sta., Bull. 47, p. 33, (1898).
do	Felt, Bull. N. Y. St. Mus., vol. VI, No. 27, p. 52, (1899).
do	Ckll. and Parrott, Industrialist, p. 281, (1899).
do	Smith, Rept. Ent., N. J. Coll. Exp. Sta., p. 446, (1899).
do	Hunter, K. U. Quart., vol. IX, No. 2, p. 104, (1899).
do	Newell, Iowa Sta., Bull. 43, pp. 170–172, (1899).
do	King, Can. Ent., vol. XXXI, p. 142, (1899).
do	Howard, Bull. No. 22, n. s., U. S. Dept. Agr., Div. Ent., p. 7, (1900).

Pulvinaria innumerabilis Felt, Bull. N. Y. St. Mus., vol. VI, No. 31, p. 581,
(1900).
do   King, Psyche, vol. IX, p. 117, (1900).
do   Brown, Bull. Wis. Nat. Hist. Soc., I, No. 1, pp. 65-
67, (1900).*
do   King, Psyche, p. 154, (1901).*
do   Hillman, Nev. Sta., Bull. 36.
do   Chambliss, Tenn. Sta. Bull., vol. X, No. 4, pp. 141-
151.
do   Piper and Doane, Wash. Sta., Bull. 36.
do   Piper, Wash. Sta., Bull. 1, pp. 121-127.

Pulvinaria pruni.

Pulvinaria pruni Hunter, K. U. Quart., vol. IX, No. 2, p. 104, (1899).

Parlatoria pergandei.

Parlatoria pergandei Comstock, Rept. Ent. in Rept. Com. Agr. 1880, p. 327.
do   Comstock, 2d Rept. Dept. Ent. Corn. Univ. Exp. Sta.,
1883, p. 113.
do   Hubbard, Ins. aff. Orange, U. S. Dept. Ent., 1885, p. 37.
do   Green, Ent. Monthly, ser. 2, 7, No. 74, p. 41, (1896).

The following synonymy is taken from C. L. Marlatt's MSS. of March 2, 1900:
Pergandei Comst. (merges into *protcus* Curt.)
Syn. var. *camelliæ* Comst.
" var. *crotonis* Ckll.
" var. *affinis* Newst.
" var. *calianthina* B. & L. (not seen; ? var. *theæ* Ckll.)
" var. *theæ* Ckll. (? *calianthina* B. & L.)
" (?) *dryandræ* Full.
" var. *euonymi* Ckll.
" *myrtus* Mask.
" (?) *pittaspori* Mask.
" *sinensis* Mask.
" var. *viridis* Ckll.
" var. *virescens* Mask.
" *viridis* Full.

Parlatoria pergandei Morgan, La. Sta., Bull. 28, 2d ser., pp. —, (18—).
La. Sta., Sp. Bull., pp. —, (18—).
Parlatoria protcus Curt., var. *pergandei* Comst., King. Can. Ent., vol. —, p. 228.
Parlatoria pergandei Reh., Zeitschrift für Entom., vol. V, p. 162, June, 1900.*
do   Ckll., Amer. Nat., July, 1897, p. 592.*
do   Craw, Rept. Bd. Hort. Cal., vol. V, pp. 41, 42, pl. VIII,
fig. 5, 1895 -'96.*

Kermes niralis.

Kermes niralis King and Ckll., Ann. and Mag. Nat. Hist., ser. 7, vol. II, 1898.
do   King, Can. Ent., p. 139, (1899), vol. XXXI.
Psyche, p. 80, July, 1900.
do   Ckll., Psyche, IX, p. 44, Apr. 1900.*

Kermes pubescens.

Kermes pubescens Bogue, Can. Ent., vol. 30, No. 7, p. 172, (1898).
    do   King, Can. Ent., p. 139.
                   Psyche, p. 80, July, 1900.
    do   Ckll, Psyche, IX, p. 44, Apr. 1900.*

Orthesia graminis.

Orthesia graminis Tinsley, Can. Ent., vol. 30, No. 1, p. 13, (1898).

NOTE.—To the list of food-plants Miss Etta Willett, a student of this department, added a number, an exact account of which was not taken at the time.

APPENDIX.

Other Coccidæ Reported from Kansas.

Kermes concinnulus Ckll., Cockerell, on oak, Can. Ent., p. 172, (1898).
Aspidiotus marlatti Parrott, on *Andropogon scoparius* and *Andropogon furcatus*, Can. Ent., p. 282, (1899).
Antonina nortoni Parrott and Ckll., on *Bouteloua racemosa*, Can. Ent., Oct. (1899).
Lecanium longulum Dougl., Parrott, Industrialist, p. 39, (1899).
Lecanium pruinosum Comst., ibid.
Aspidiotus cyanophylli Sign., ibid.
Aulacaspis boisduvalii Sign., ibid.
Parlatoria proteus Curt., ibid.
Aspidiotus helianthi Parrott, Can. Ent., vol. 31, p. 176 (1899).
Antonina bouteloue Parrott, on *Bouteloua hirsuta*, Parrott, Kan. Agr. Coll. Bull. No. 98, p. 138, (1900).
Antonina purpurea Sign., on *Milium* and *Agripyrum*, ibid.
Antonina graminis Parrott, on *Eragrotis trichodes*, *Bulbilis dactyloides*, *Paspalum ciliatifolium*, ibid., p. 140.
Gymnococus nativus Parrott, on *Sporobolus cryptandrus*, ibid., p. 143.
Pseudolecanium obscurum Parrott, on *Androgopon scoparius* and *Sporobolis longifolius*, ibid., p. 145.
Pseudolecanium californicum Ehrhorn, on *Androgopon furcatus*, ibid., p. 145.
Ericoccus kemptonia Parrott, ibid., p. 144.
Pulvinaria hunteri, on maple, King, MS.

LABORATORY OF COMPARATIVE ZOOLOGY AND ENTOMOLOGY,
    July, 1901.

GLOSSARY.

Abdomen—The part of the insect posterior to the thorax.

Anal lobes or plates—A pair of small triangular hinged processes forming a valve covering the anal orifice in the Lecaniinæ.

Anal orifice—The external opening of the intestine visible as a circular opening in the central part of the pygidium.

Anal ring—A chitinous ring around the anal orifice.

Anal tubercles—A pair of rounded or conical processes found one on each side of the anal orifice in the larva of Hemicoccinæ.

Antennæ—A pair of jointed sensory organs used as feelers, found on the head.

Appendages—Outgrowths from the body or any part of the body; used as a general term for mouth-parts, antennæ, legs and wings, or processes on the tarsi called digitules.

Apterous—Wingless.

Balancers—(See halteres).

Carina—A keel or ridge.

Carinated—Having a keel or ridge.

Caudad—Towards the caudal or posterior part of the body.

Caudal—Pertaining to the posterior extremity.

Cephalic—Pertaining to the anterior or head end of the body.

Cephalad—Towards the head.

Chitin—A hard, tough, horny substance found in the skin and hard parts of insects.

Circumgenital glands—(Also called grouped glands, ventral glands and spinnerets). Small circular glands arranged round the genital orifice in groups resembling bunches of grapes.

Compressed—Flattened laterally.

Coxa—Basal joint of the leg.

Cuticle—Thin outer skin.

Depressed—Flattened from above.

Digitules—Appendages often found on the feet of coccidæ. Found in Lecanium as knobbed hairs.

Dorsad—Towards the back.

Dorsal—Pertaining to the back of the insect.

Dorsal scale—The part of the scale which covers insects belonging to the Diaspinæ.

Ecdysis—The moult or change of skin.

Exuviæ—The cast-off skins, forming part of the dorsal scale.

Femur—The third joint of the leg, between the trochanter and the tibia, but for measuring purposes the trochanter and femur being fused together are considered as one piece.

Filiform—Thread-shaped.

Grouped glands—(See circumgenital glands).

Halteres, also called balancers—A pair of small organs replacing the hind wings in the males of Coccidæ. They have a strap-shaped basal part with one or more stout hooked bristles on the end.

Honey-dew—A sticky substance secreted by Coccidæ and some other Homoptera.

Incisions—Marginal slits or notches.

Laterad—Towards the side, away from the median line.

Larva, larval stages—Stages of the insect up to the pupa, in Diaspinæ to the second moult.

Lobe—A prominent rounded process found on the caudal margin of the pygidium in the Diaspinæ.

Mesad—Toward the median line.

Mesal—Relating to the middle.

Metamorphosis—Change of form during development.

Millimetre (mm.)—The 1000th part of a metre; approximately equal to the twenty-fifth part of an inch.

Oviparous—Laying eggs.

Oviposition—The act of laying eggs.

Parasitised—Affected by parasites.

Parthenogenesis—Reproduction by the female without intercourse with the male, by internal budding.

Pellicles—Exuviæ.

Plates, or squames—Broad, flat, horny, transparent organs found along the posterior and lateral edges of the body of some of the Coccidæ.

Process—Any noticeable projection or extension of the body.

Puparium—Covering scale of the Diaspinæ.

Pupa—The chrysalis.

Pygidium—Caudal portion of the abdomen of the Diaspinæ formed by the fusion of a number of segments.

Sac—The cottony covering secreted by some Coccidæ.

Scale—The covering of the Diaspinæ, formed of cast-off skins and excretions, also the waxy covering of the male Lecanid. The word is often used as a general term for any insect belonging to the family of Coccidæ whether they produce a scale or not.

Secretion—Matter produced by the various glands of the body. It may be made of closely-woven fibers as in the Diaspinæ, or be waxy, cottony, or mealy.

Secretory—Concerned in the process of secretion.

Segments, Somites—The transverse divisions of the body.

Seta—A bristle or long, stiff hair.

Setiferous—Bearing setæ.

Somites—(See segments).

Spatulate—Flattened and expanded at the tip.

Spinnerets—The secretory organs found in various parts of the body, consisting of internal tubes terminating in pores, spines or conical hairs. In the Diaspinæ there are found groups of spinnerets in the pygidium, called ventral, grouped, or circumgenital glands.

Spiracles—External openings of the trachew.

Squames—(See plates).

Suctorial—Having sucking mouth-parts.

Tarsi—The distal joints of the leg succeeding the tibia.

Tarsal—Pertaining to the tarsi.

Test—The waxy, glassy, or horny covering of various Coccidæ.

Tibia—The joint of the leg between the femur and the tarsi.

Trochanter—The joint between the femur and the coxa, and usually fused with the femur.

Truncate—Having the appearance of having the extremity cut off by a plane parallel to the base.

Ventral—Pertaining to the under side of the insect.

Ventral scale—The part of the scale under the insect.

Wax glands—Glands secreting wax, found in the pygidium as circumgenital glands; also in other parts of the body, but of different structure.

INDEX.

(79)

PLATE I.

Fig. 1. *Aspidiotus forbesi* Johns. One side of anal plate of female. On cherry. Lawrence.

Fig. 2. *Aspidiotus forbesi* Johns. Anal plate of female. On crab apple. Lawrence.

Fig. 3. Variations in anal plate of the *forbesi* found on crab apple and illustrated in Fig. 2.

Fig. 4. *Aspidiotus forbesi* Johns. Anal plate of female. On apple. Lawrence.

Fig. 5. *Aspidiotus forbesi* Johns. Anal plate of female. On cherry. Seward Co., Kansas.

Fig. 1.

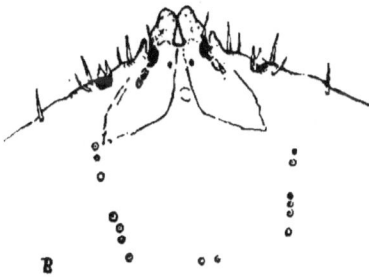

Fig. 2.

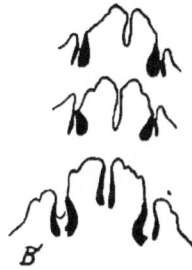

Fig. 3.

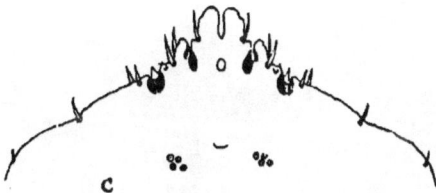

Fig. 4.

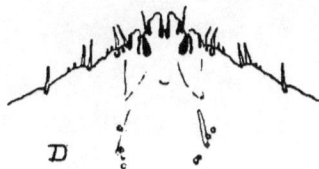

Fig. 5.

Ella Weeks and S. J. Hunter, ad nat. del.

PLATE II.

Fig. 6. *Aspidiotus forbesi* Johns. Anal plate of female. On apple. Lawrence.

Fig. 7. *Aspidiotus forbesi* Johns. Anal plate of female. On cherry. Two miles N. W. of Lawrence.

The letters upon these Figures have no reference to letters distinguishing lots in the text upon the species. They are placed upon the Figures to call attention to the series, which are produced to illustrate variations and resemblances of *forbesi* brought together from widely separated localities and taken from different hosts.

Fig. 8. *Aspidiotus ancyclus* Putnam. Anal plate of female. On maple. Lawrence.

Fig. 9. *Aspidiotus ancyclus* Putnam. Anal plate of female. On maple. Chitinous processes deeper, more club-shaped than typical specimen. Dorsal glands smaller but more numerous than in lot from which Figure 8 was drawn.

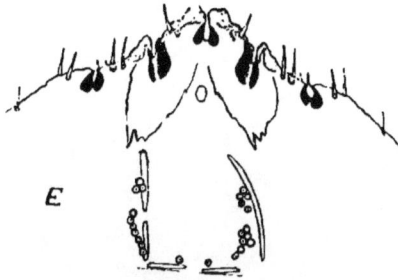

Fig. 6.

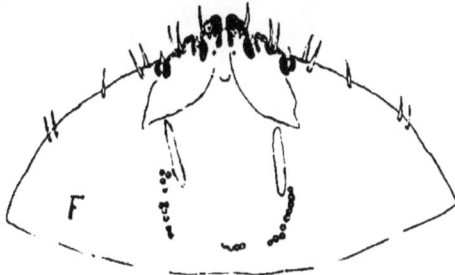

Fig. 7.

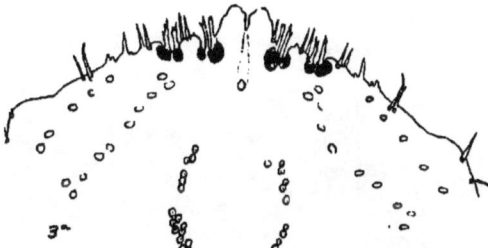

Fig. 8.

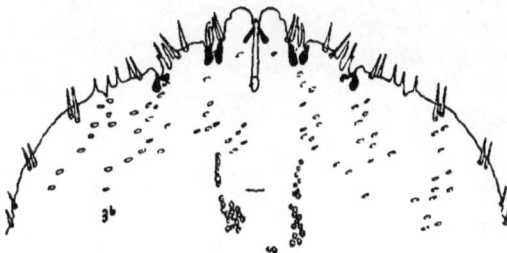

Fig. 9

Ella Weeks and S. J. Hunter, ad nat. del.

PLATE III.

Fig. 10. *Aspidiotus uvœ* Comst. Anal plate of female. On grape. Iola.

Fig. 11. *Aspidiotus uvœ* Comst. var. anal plate of female. On *Carya abla* Nutt. Lawrence.

Fig. 12. *Aspidiotus osborni* Newell and Ckll. Anal plate of female.

Fig. 13. Antenna, cephalic leg, and caudal portion of body of Nymph of *Aspidiotus osborni* Newell and Ckll.

Fig. 14. *Aspidiotus ulmi* Johns. Anal plate of female.

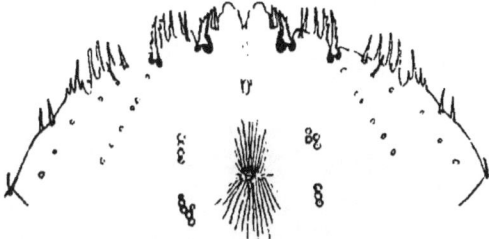

Fig. 10.

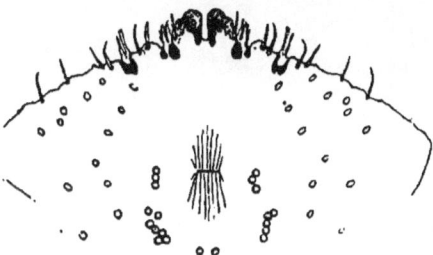

Fig. 11

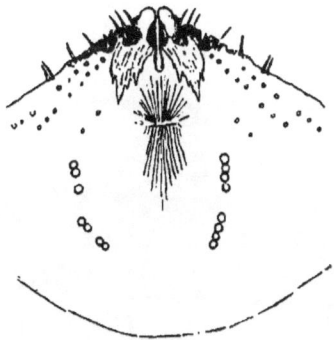

Fig. 12.

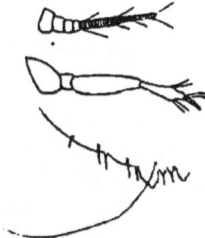

Fig. 13.

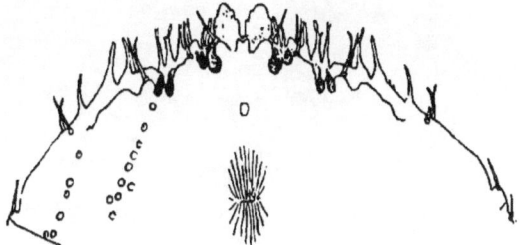

Fig 14.

Ella Weeks and S. J. Hunter, ad nat. del.

PLATE IV.

Fig. 15. *Aspidiotus fernaldi* Ckll., sub species *albiventer* sub sp. nov. Anal plate of female.

Fig. 16. Variations in anal plate of female of sub. sp. *albiventer*.

Fig. 17. *Aspidiotus juglans-regiæ* Comst. var. anal plate of female.

Fig. 18. *Aspidiotus perniciosus* Comst. San Jose scale. Anal plate of female, showing variation from original description.

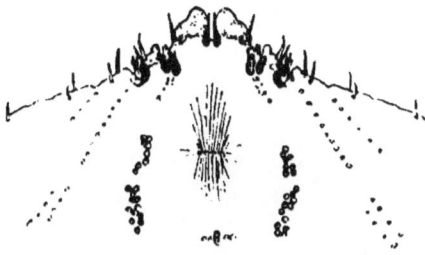

Fig. 15.

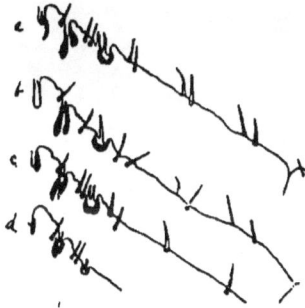

Fig. 16.

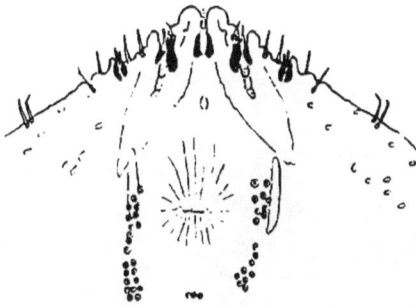

Fig. 17.

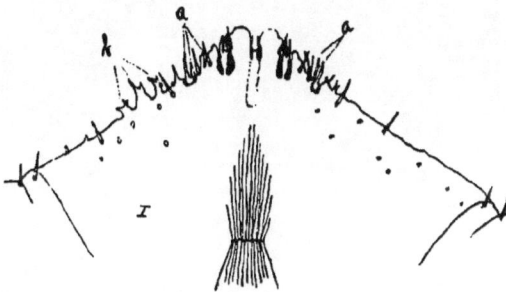

Fig. 18.

Ella Weeks and S. J. Hunter, ad nat. del.

PLATE V.

Fig. 19. *Aspidiotus perniciosus* Comst. Anal plate of female showing normal structure of this lot of specimens.

Fig. 20. *Aspidiotus perniciosus.* The upper figure is the left side of anal plate and the lower figure is the right side of another specimen transposed for comparison.

Figs. 21, 22. *Aspidiotus perniciosus.* One side of anal plate of females showing variations.

Fig. 23. Newly born nymph of *A. perniciosus* Comst. Antenna and cephalic leg enlarged on the right.

I

Fig. 19.

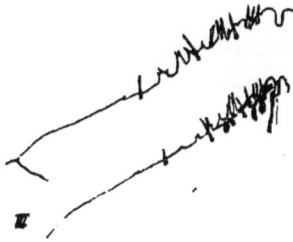

II

Fig. 20.

a

Fig. 21.

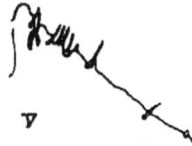

V

Fig. 22.

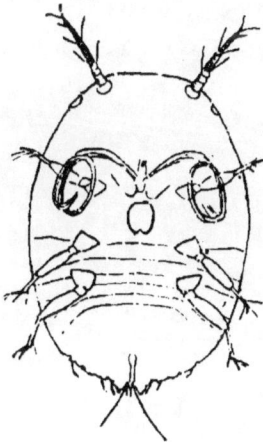

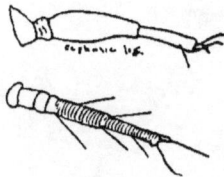

Fig. 23.

Ella Weets and S. J. Hunter, ad nat. del.

PLATE VI.

Fig. 24.   *Aspidiotus greenii* Ckll.   Anal plate of female.

Fig. 25.   *Aspidiotus obscurus* Comst.   Anal plate of female.

Fig. 26.   *Aspidiotus hederæ* Vall., var. *nerii* Bouche'.   Male scales showing circular shape and central position of exuviæ.

Fig. 27.   *Aspidiotus hederæ* Vall., var. *nerii* Bouche'.   Anal plate of female.

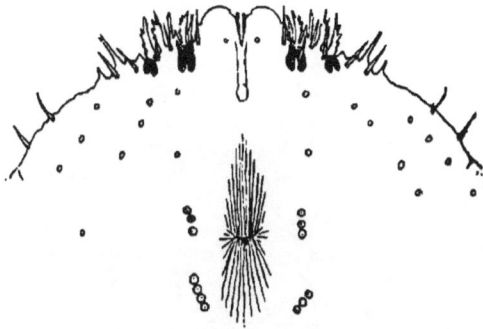

Fig. 24.

Fig. 25.

Fig. 26.

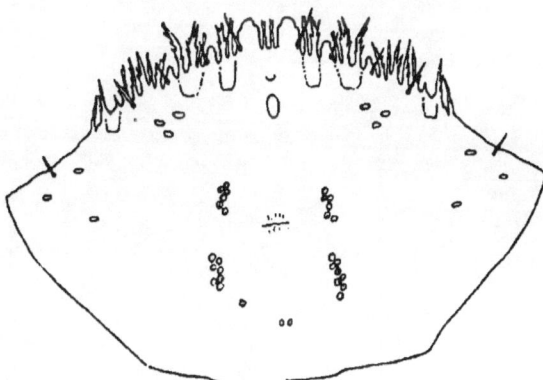

Fig. 27.

Ella Weeks and S. J. Hunter, ad nat. del.

PLATE VII.

Fig. 28. Wing, antenna, and cephalic leg of male of *A.* v. *nerii* Bouche'.

Fig. 29. Sketch showing variations in shape of the three pairs of lobes in *A.* v. *nerii* Bouche'.

Fig. 30. *Aspidiotus æsculi* Johns., sub species *solus*, sub. sp. nov. A median group of from one to two circumgenital glands was found in majority of the specimens examined. They are not shown in this figure, as they are frequently concealed beneath the club here shown. This sketch further illustrates the extreme number of plates found.

Fig. 31. *Diaspis snowii* sp. nov. Anal plate of female.

Fig. 32. *Diaspis snowii* sp. nov. Scales of males.

Fig. 33. *Diaspis snowii* sp. nov. Scales of females.

Fig. 28. Fig. 29.

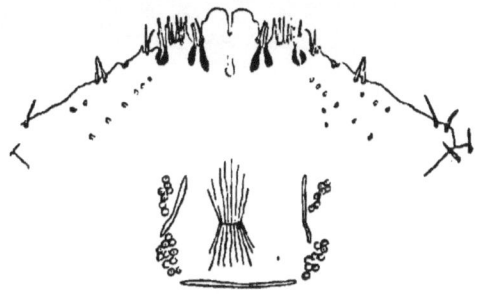

Fig. 30.

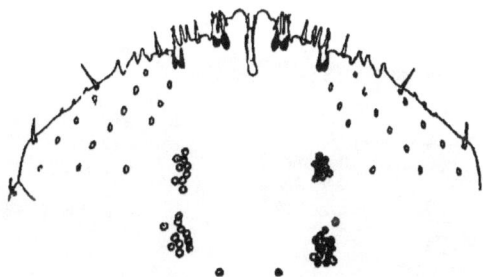

Fig. 31

Fig. 32. Fig. 33.

Ella Weeks and S. J. Hunter, ad nat. del.

PLATE VIII.

Fig. 1. *Lecanium maclurœ* nov. sp. on osage orange. Adult female on center of twig, immature female on side of twig. Enlarged six times.

Fig. 2. *Lecanium maclurœ* nov. sp. Antenna, prothoracic leg, mesothoracic leg, and metathoracic leg of adult female. Claws of pro- and mesothoracic legs enlarged beneath respective members. Two digitules of claw seen only in mesothoracic leg. Enlarged about 125 times.

Fig. 1.

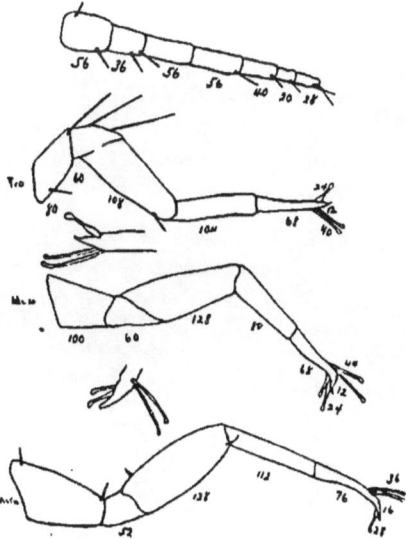

Fig. 2.

PLATE IX.

Fig. 1. *Lecanium canadense*, Ckll. on *Ulmus americana.* Dorsal view of female. Enlarged ten times.

Fig. 2. *Lecanium canadense*, Ckll. on *Ulmus americana.* Lateral view of female. Enlarged ten times.

Fig. 3. *Lecanium canadense*, Ckll. Antenna, and metathoracic leg of female. Enlarged about 125 times.

Fig. 4. *Lecanium kansasense*, nov. sp. Dorsal view of female, on *Cercis canadensis.* Enlarged six times.

Fig. 5. *Lecanium kansasense.* Antenna and leg of female. Enlarged about 125 times.

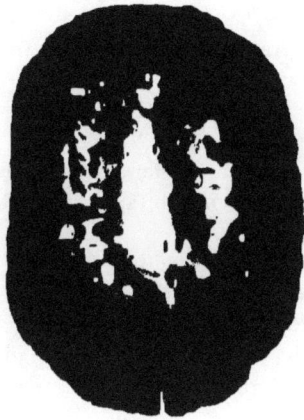

Fig. 1.

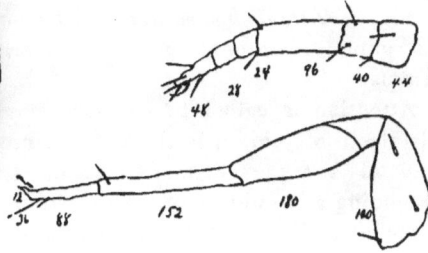

Fig. 3.

Fig. 2.

Fig. 4.

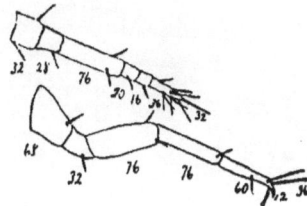

Fig. 5.

Ella Weeks, del. ad nat.

PLATE X.

Fig. 1. *Lecanium cockerelli*, nov. sp. Dorsal view of female on *Ulmus americana*. Enlarged 12 times.

Fig. 2. *Lecanium cockerelli*, nov. sp. Antenna and leg of female on *Ulmus americana*. Enlarged about 125 times.

Fig. 3. *Lecanium cockerelli*, nov. sp. Antenna and leg of female on *Juglans nigra*. Enlarged about 125 times.

Fig. 4. *Lecanium armeniacum* Craw. *a*. Antenna of female bearing six joints; *b*, antenna of female bearing seven joints; *c*, leg of female.

Attention is called to the fact that bristle extending from the distal end of joint 3, in the seven-jointed variety, was not found in any of the 13 specimens examined of the six-jointed variety, extending as would be expected from the middle of the long 3d segment. The long 3d segment always bore hairs at distal end and nowhere else.

Fig 1.

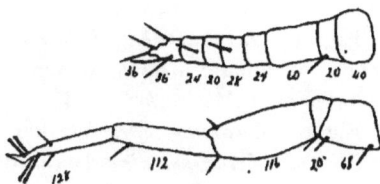

Fig 2.

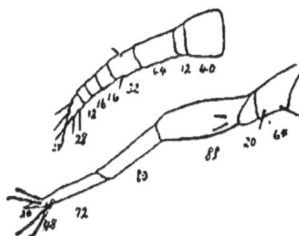

Fig. 3.

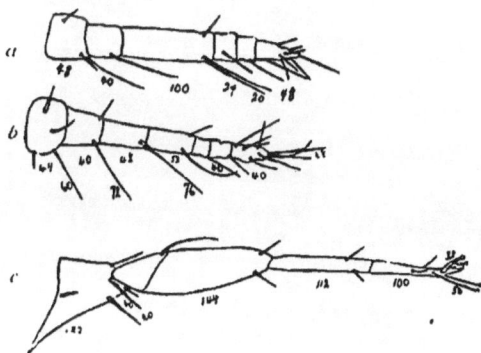

Fig. 4.

Ella Weeks del. ad nat.

PLATE XI.

Fig. 1. *Lecanium hesperidum* L. Dorsal view of female on *Citrus* sp. Enlarged 15 times.

Fig. 2. *Lecanium hesperidum* L. *a* Antenna of female on *Citrus* sp. Enlarged about 125 times. *b.* Antenna of female as delineated by Berlese (Tracing sent by Professor Cockerell).

Fig. 3. *Lecanium hesperidum* L. *a* Antenna of female. *b.* Antenna of female drawn to show difference in sizes. *c.* Leg of female. On *Hedera helix.* Enlarged about 125 times.

Fig 4. *Lecanium coffeæ* Walker. Antenna and leg of female.

Fig. 5. *Lecanium oleæ* Bernard. Antenna and leg of female, 5th joint not always distinct. On *Nerium oleander.* Enlarged about 125 times.

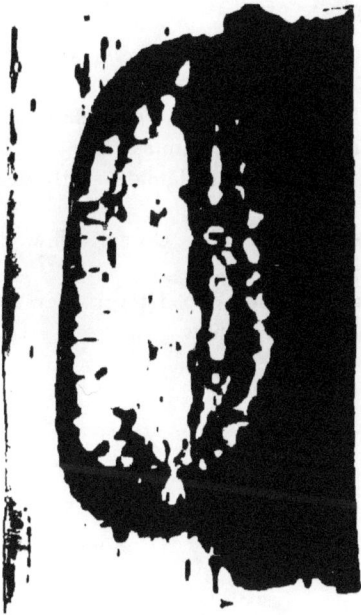

Fig. 1.

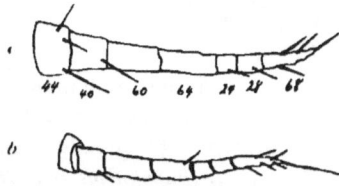

Fig. 2.

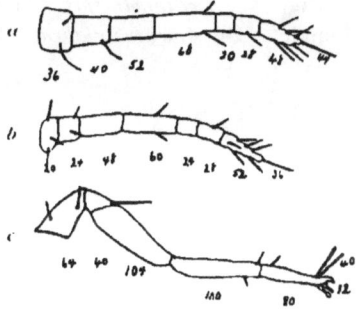

Fig. 3.

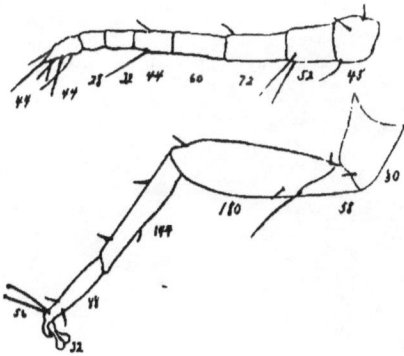

Fig. 4.

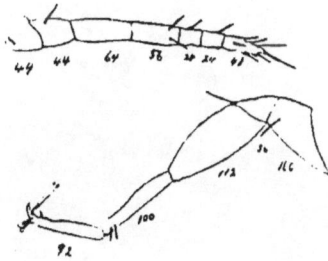

Fig. 5.

PLATE XII.

Fig. 1. *Lecaniodiaspis celtidis* Ckll , sub. sp. *pruinosus* sub. sp. nov. Dorsal view of female on *Ulmus americana*.

Fig. 2. *Lecaniodiaspis celtidis* Ckll. sub. sp. *pruinosus* sub. sp. nov. Antenna of female. Enlarged about 125 times.

Fig. 3. *Lecaniodiaspis celtidis* Ckll. Antenna of female. Enlarged about 125 times.

Fig. 4. *Lecaniodiaspis* (?) *parrotti*, nov. sp. Dorsal view of female on *Aesculus glabra*. Enlarged 12 times.

Fig. 5. *Lecaniodiaspis* (?) *parrotti*, nov. sp. lateral view of female. On *Aesculus glabra*. Enlarged 12 times.

Fig. 1.

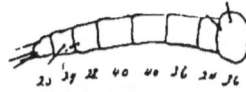

Fig. 2.

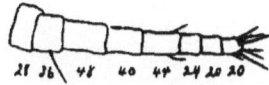

Fig. 3.

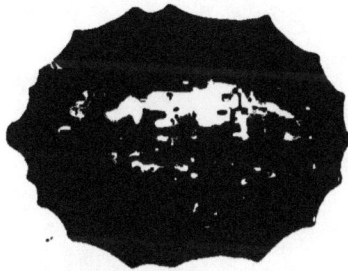

Fig. 4.

Fig. 5.

PLATE XIII.

Fig. 1.—*Chionaspis ortholobis* Comstock, on *Salix* sp. Anal plate of female, showing somewhat irregular position of dorsal glands.

Fig. 2.—*Chionaspis ortholobis* Comstock, on cottonwood, *Populus* sp. Anal plate of female, illustrating variations in second and third lobes. The dorsal glands, second and third rows, in this figure, as in figure 1, are shown beneath the circumgenital glands. These are frequently located laterad of circumgenital glands.

PLATE XIII.

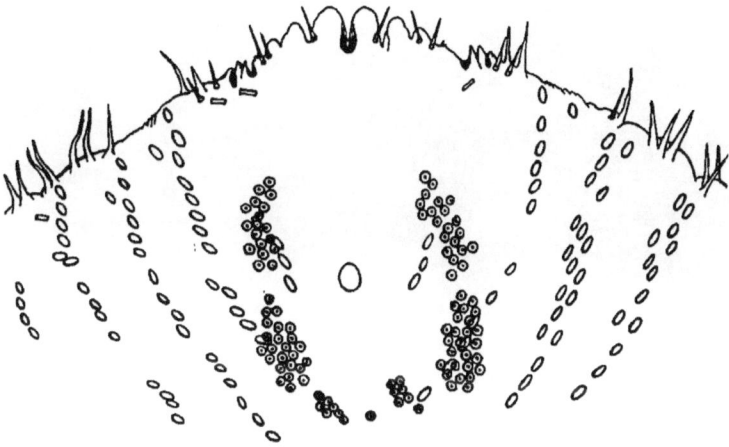

Fig. 1.

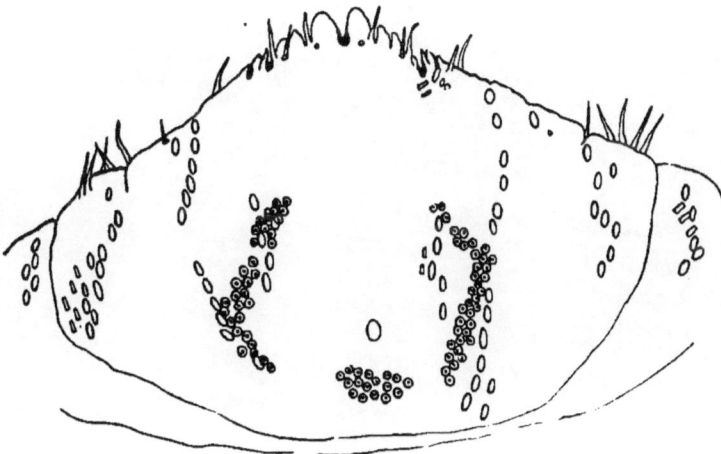

Ella Weeks, del. ad nat.   Fig. 2.

PLATE XIV.

Fig. 1.—*Chionaspis salicis-nigræ* Walsh, on *Salix* sp. Anal plate of female. '

Fig. 2.—*Chionaspis americana* Johnson, on *Ulmus americana*. Anal plate of female. (*a*) Variation in margin of median and second lobe; (*b*) illustrates marginal variations and forked plates. '

PLATE XIV.

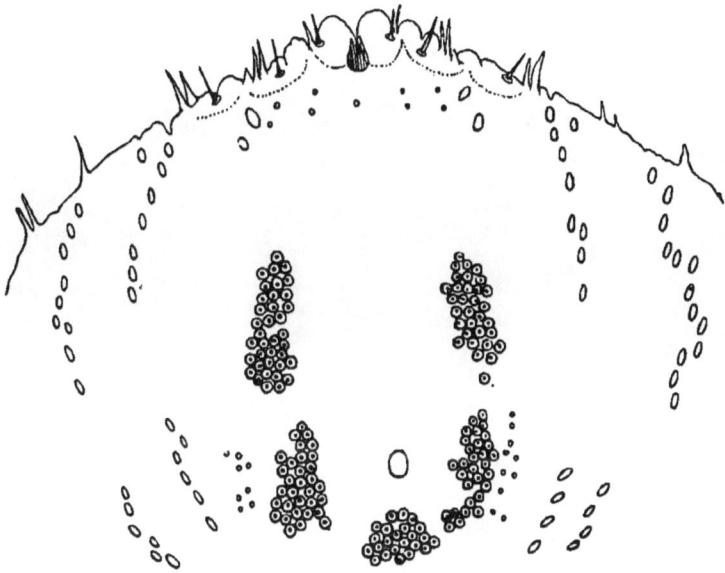

Fig. 1.

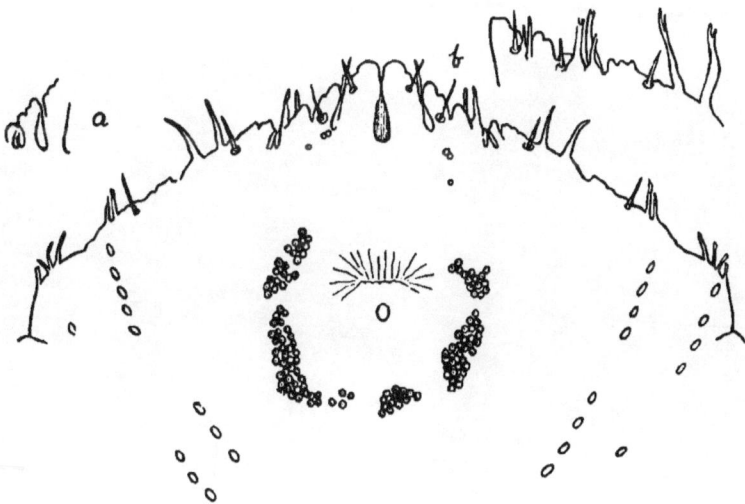

Ella Weeks, del. ad nat.   Fig. 2.

PLATE XV.

Fig. 1.—*Chionaspis platani* Cooley, on sycamore, *Platanus occidentalis*. Anal plate of female.

Fig. 2.—*Chionaspis pinifoliæ* Fitch, on *Pinus* sp.   Anal plate of female.

PLATE XV.

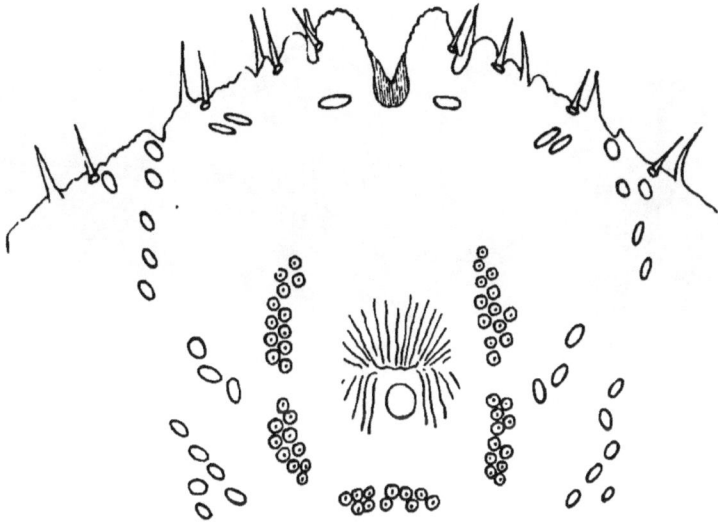

Fig. 1.

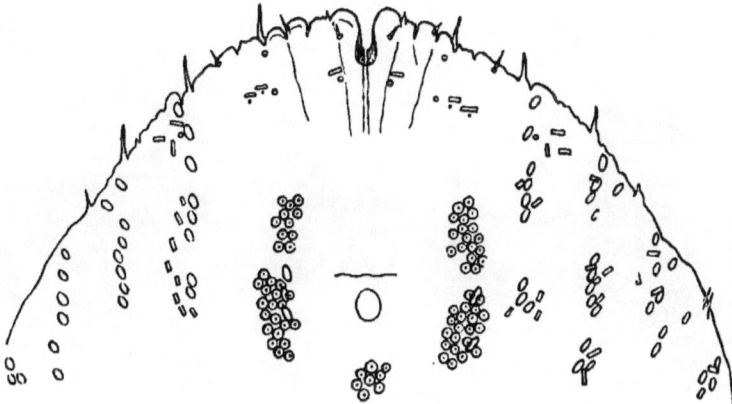

Ella Weeks, del. ad nat.   Fig. 2.

PLATE XVI.

Fig. 1.—*Pulvinaria innumerabilis* Rathv. Leg and antenna of female. Greatly enlarged.

Fig. 2.—*Pulvinaria innumerabilis* Rathv. Adult female on twig, the cottony covering of the egg mass frayed by the weather. Enlarged.

Fig. 3.—*Pulvinaria pruni*, n. sp. Sketch of nymph, its antenna, and leg, at time of location upon the plum leaf. Enlarged.

PLATE XVI.

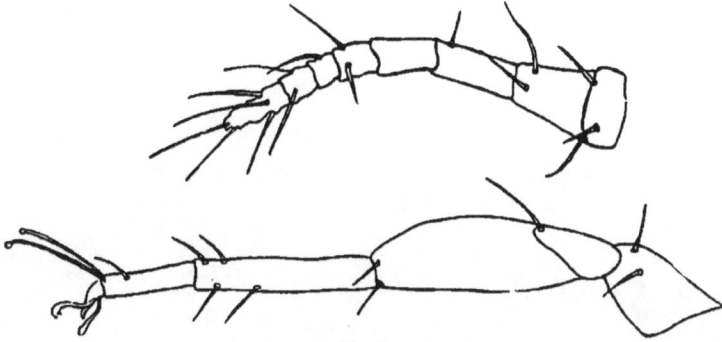

Fig. 1.

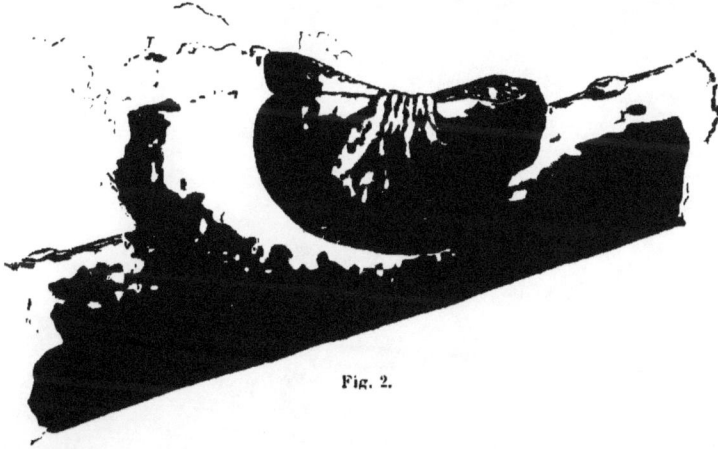

Fig. 2.

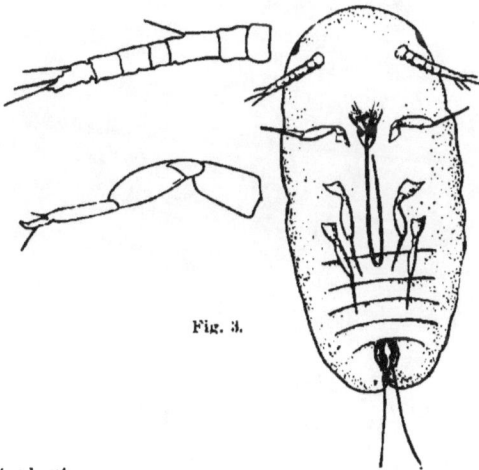

Fig. 3.

Ella Weeks, del. ad nat.

PLATE XVII.

Pulvinaria pruni, n. sp., on leaves and twig of plum. Adult females, being much recurved and shriveled, do not show clearly at the ends of the cottony egg masses.

PLATE XVII.

PLATE XVIII.

Fig. 1.—*Pulvinaria pruni*, n. sp. Antenna. Anterior (*A*), median (*B*), posterior (*C*) legs of adult female on leaves of plum. Greatly enlarged.

Fig. 2.—Antenna. Anterior (*A*), median (*B*), posterior (*C*) legs of adult female on plum twigs. Greatly enlarged.

PLATE XVIII.

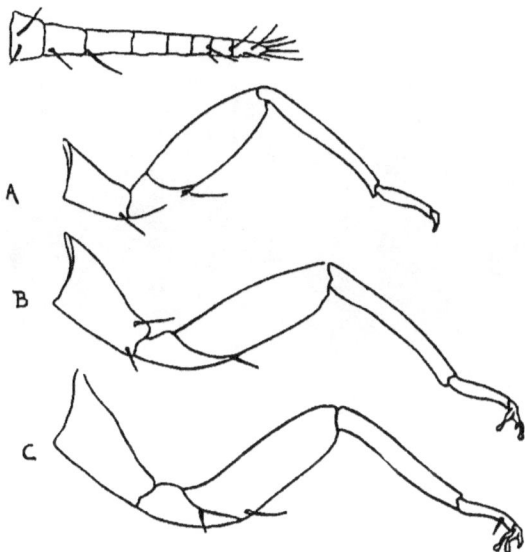

Fig. 1.

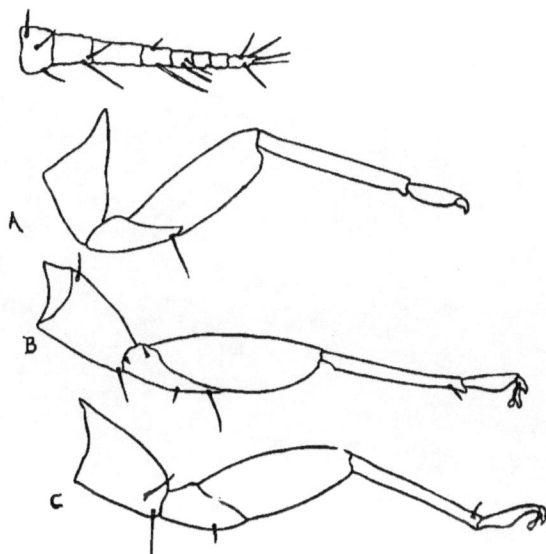

Ella Weeks, del. ad nat.   Fig. 2.

PLATE XIX.

Fig. 1.—*Parlatoria pergandei* Comstock, on orange, *Citrus* sp.

Fig. 2.—(a) *Parlatoria pergandei*. First, second, third and fourth lobes, with variations. (b) *Parlatoria proteus* Curt., on *Pinus insignis*.

Note the variation in the papillar fourth lobe of *pergandei*, and presence of plate beyond fourth lobe of *proteus*.

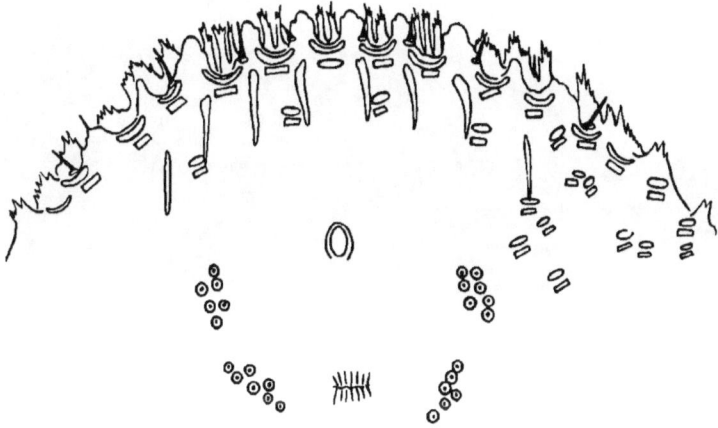

Fig. 1.

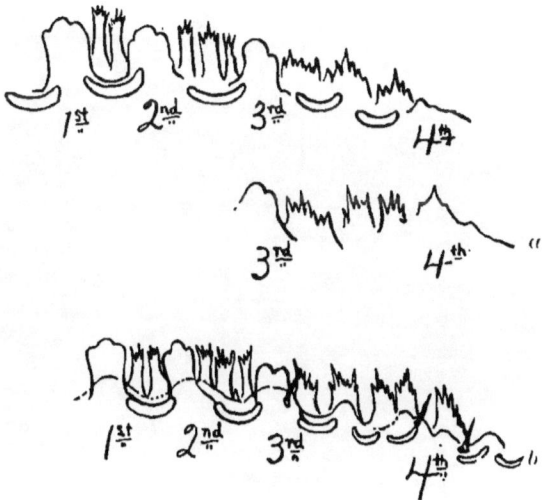

Ella Weeks, del. ad nat.   Fig. 2.

FIG. 1. *Kermes pubescens* on *Quercus alba*, Lawrence.

FIG. 2.—*Kermes nivalis* on *Quercus alba*, Lawrence.

FIG. 3.- -Ventral view of *Orthezia graminis* on goldenrod (*Solidago* sp.), without posteriorly elongated egg sac. Mrs. S. G. Cady, col., Blue Rapids, Marshall county.

FIG. 4.—Dorsal view of *Orthezia graminis*, showing posteriorly elongated egg sac.

Photographs by the author.